CHASING
THE
MOLECULE

CHASING
THE
MOLECULE

JOHN BUCKINGHAM

SUTTON PUBLISHING

First published in the United Kingdom in 2004 by
Sutton Publishing Limited · Phoenix Mill
Thrupp · Stroud · Gloucestershire · GL5 2BU

British Library Cataloguing in Publication Data
A catalogue record for this book is available from the British Library.

ISBN 0-7509-3345-3

Typeset in 11/14.5pt Sabon.
Typesetting and origination by
Sutton Publishing Limited.
Printed and bound in England by
J.H. Haynes & Co. Ltd, Sparkford.

To Briar

Contents

List of Plates

Picture Credits

British Library (Plate 21)
Justus von Liebig University, Giessen (Plate 4)
Linnean Society (Plate 2)
Royal Pharmaceutical Society (Plates 3, 7)
Royal Society of Chemistry (Plates 5, 6, 9, 10, 11, 12, 13, 14, 15, 16, 17)
Website Kekules Traum (Plate 20)
Wellcome Library (Plates 1, 8, 18, 19)

The caption to Plate 19 is adapted from Alexander Roob, *The Hermetic Museum: Alchemy and Mysticism*, Taschen Books, 1996, p. 402.

Acknowledgements

I am grateful to Professor Bernard Aylett and to Carol Mellor for reading large sections of the draft manuscript and for offering invaluable advice on both factual and stylistic points, and for helping with the glossary. Any remaining inaccuracies remain my own responsibility. Thanks also go to John Thompson, Frank Hopkins, Iwona Sakowicz and the staff of the various research libraries in London, including the Wellcome Library, the British Library, the London Library and the libraries of the Royal Society of Chemistry, the Royal Pharmaceutical Society and the Linnean Society; also to the staff of Sutton Publishing, not least to Peter Clifford, for suggesting the book's title, Jaqueline Mitchell and Hilary Walford. Dr Matt Griffiths drew the chemical formulae and other line diagrams.

Note on Molecular Formulae

In the narrative I have made no attempt to give a rigorous description of how different chemists in the nineteenth century arrived at their various atomic weights and formulae. The derivation of 'two-volume' and 'four-volume' formulae is complex. (For a more detailed explanation, see Rocke 1984, p. 54.) Incorrect formulae are shown throughout using superscripts, and their correct modern equivalents using modern subscripts. Thus benzoic acid according to Liebig was $C^{14}H^{12}O^4$; the correct formula is $C_7H_6O_2$.

This does not correspond exactly with the historical record. The change from superscripts to subscripts was first made by Liebig in 1834, but his formulae continued to be wrong after that date. On the other hand, most French chemists continued to use superscripts until the end of the century, although by then their formulae would have been correct.

Introduction

On 11 March 1890, a five-hour banquet for hundreds of invited guests was held in the festive chamber of the Berlin City Hall, 'A festival of a magnificence perhaps unparalleled in the history of science', as one (English) commentator later wrote.[1] The vast chandeliered room was decorated with palm trees and laurel leaves, and one end was dominated by a five-metre-high oil painting of Bismarck and other European statesmen carving up the Turkish empire at the Congress of Berlin. Invitees were obliged to wear white tie and tails, and the occasion was considered so important that lady guests were admitted, to sit at separate tables in an arcade. His Imperial Majesty Kaiser Wilhelm II was invited, but sent his regrets and was represented by the Minister of Education. The Kaiser was deeply embroiled in disagreements with Bismarck, whom he sacked only nine days later, and was also being blackmailed by a prostitute.[2]

The first speaker at the banquet was A.W. Hofmann, one of the most distinguished scientists of his era and the foremost student of the famous Justus von Liebig, who had effectively founded the science of organic chemistry seventy years before. Now seventy-one and near the end of his life, Hofmann gave a light-hearted address. Nine more speeches followed. During one of them, the president of the Chemical Society of London showed the audience a small tube containing a colourless liquid, which he had brought with him from Engand, and which had been sealed more than sixty years before by its discoverer, Michael Faraday. Another speech was a review of the current state of chemistry by Adolf von Baeyer, the discoverer of barbiturates, but the rest of the contributions were congratulations from all over the world on the occasion of the twenty-fifth anniversary of the great achievement of the man who rose to speak last.

No verbatim account was made of exactly what he said. We have to rely on various accounts written down subsequently, days or years later. It is not even certain that he said what he is supposed to have said. He may have invented most of it when he was asked some days later to prepare a written version for publication.

Some time after the banquet was over, the young chemist charged with putting together the published report sent a draft to Heinrich Seidel, the popular novelist, asking him if he could be so kind as to suggest any necessary stylistic improvements. The reply that he received was not what he expected: 'I read the report with interest, especially the speech of the guest of honour. I have to admit that never before have I come across such an example of self-adulation and conceit. This is perhaps only possible in academic circles . . .'.[3]

The important scientific breakthrough of the 1860s that was being celebrated at the banquet came on the heels of several decades of intellectual struggle, with many blind alleys, but when it took place it rapidly led to the foundation of whole new industries for both peace and war. The twentieth century, for good or ill, was the century of chemistry. Just as now we are beginning to reap the real benefits of the DNA structure breakthrough of the 1950s and the development of computing from about the same era, so the industries of the twentieth century grew at an accelerating pace from the chemical insights of the 1860s. These in turn were founded on the study of natural products garnered from the four corners of the world.

The science of natural products, or organic chemistry, was from an early date recognized as being the chemistry of one vital element, carbon. How and why just one element from among the dozens known even by the beginning of the nineteenth century should be the foundation stone of the staggering diversity of nature's products seemed completely inexplicable. It was no wonder that nearly all scientists of the time believed that chemistry alone was not enough, and that there must be some kind of divinely inspired vital force. It took more than half a century of painstaking work for the foundation stones of the 1865 breakthrough to be laid. True organic chemistry began with the researches of Liebig and Wöhler in Germany, and Berzelius in Sweden, and proceeded through the remarkable trio of French chemists Dumas, Laurent and Gerhardt.

The interplay between these strong personalities in a Europe that was full of national and personal rivalry but not yet as nationalistically riven as it was to become later, takes us back to an age when, in science, everything seemed possible. The results of these early organic chemists were rendered uncertain by some fundamental and, to us, rather trivial misunderstandings. The unravelling of these uncertainties took the involvement of a number of British chemists of the 1850s, and then of the brilliant Italian, Cannizzaro, who emphasized the importance of the overlooked work of his countryman, Avogadro, half a century before. The time was ripe for a monumental change of insight, but the way in which this emerged was more complex than the speeches at the banquet might have us believe.

As well as a similarity between the events of the 1860s celebrated at the banquet, and those at Cambridge ninety years later, there is a direct link. Everyone has seen photographs of Watson and Crick posing beside their DNA model. Their assembly of metal rods and pieces of tin explaining the properties of the molecules of life could not have taken place without the chemical bonding ideas that sprang into being nearly a century before. The history of that earlier breakthrough is just as mired in controversy as the events of the 1950s, but, because it is a century further back in time with all the protagonists long dead, it is not so well known to the modern reader. Much has been written about whether Rosalind Franklin was unjustly denied credit for her DNA work, but at least it has never been suggested that in order to resolve the question it is necessary to enter the realm of dreams, which is something that we shall have to do in due course before we can try to decide whether the adulation at the banquet was justified.

The breakthrough was centred on the belated realization that there was such a thing as a molecule, the smallest particle of a chemical compound that can exist, and that the molecule has a shape and size and properties, even though it is far too small to be seen. It was the absence of a clear concept of the molecule that led science down its blind channels, and it was a major feat of the human intellect and imagination to guide it out again. The question of how credit for this helmsmanship should be apportioned is one that will never be satisfactorily resolved. Many were involved, and one, August Kekulé, the guest of honour at the banquet, has been bitterly accused of unjustly appropriating the achievements of others.

This book is an attempt to give an accessible account of these events of the 1860s and the decades leading up to them. In following this story we will meet some of science's most flamboyant, and some of its most tragic, figures, some of whose stories are known only hazily, or not at all, to non-chemists. In order to put these happenings into perspective, we must first make a brief journey back to the Middle Ages and beyond.

ONE

A Leaf is not a Stone

Before discovering chemistry, it was necessary to discover the art of discovering chemistry.
                    S. Brown (1843), cited in Mellor (1922: 18)

Well over a century before the Berlin banquet, the famous lexicographer and sage Dr Samuel Johnson came out of a church with his friend and biographer James Boswell. It was Saturday, 6 August 1763, and he had accompanied Boswell to Harwich in the east of England, where Boswell was due to take ship for the Netherlands. The two spent time together looking round the town. As they left the church, they fell to discussing Bishop Berkeley's 'Ingenious sophistry to prove the non-existence of matter, and that every thing in the universe is merely ideal'. Boswell observed that, although everyone was satisfied that the doctrine was not true, it was impossible to refute it. Recalling the occasion, he wrote, 'I shall never forget the alacrity with which Johnson answered, striking his foot with mighty force against a large stone, till he rebounded from it. "I refute it *thus*," he said.'[1]

Some years later, as Boswell relates, Johnson came up with a much more sardonic riposte to a visitor who 'Thought fit to maintain Dr Berkeley's ingenious philosophy'. As the man was leaving, Johnson said to him, 'Pray, sir, don't leave us: for we may perhaps forget to think of you, and then you will cease to exist.'[2]

The existence of matter is the axiom on which all useful knowledge rests. Since the earliest times, people have wondered about the nature of reality and the relationship of the things around them to the inner world of the spirit. We cannot enquire about this inner nature without being here in the material world. We need our brain and the mouth with which to feed it, before we can start to wonder about the nature of any spirit life that may or may not exist. Even Socrates ceased to think after he was poisoned, or at the very least was unable to communicate the results of what had most recently happened to him.

Assuming then that Johnson, who was very interested in chemistry and for many years carried out experiments himself,[3] was right and that matter truly does exist, more intricate questions arise. What *is* a stone, what *is* a leaf, why is one inert while the other grows? What is the difference between the 'inorganic' stone and the 'organic' leaf? Are such questions even answerable, and what role, if anything, does divine spirit play in giving life to the leaf? To people in earlier centuries, the answers to such questions were far from self-evident, except to those who were devoutly and unthinkingly religious.

Chemistry is no more and no less than the science of the different kinds of matter. To the chemist, current-day talk about food or other materials being 'free from chemicals' is ludicrous; *everything* is chemicals; there are only good chemicals and bad ones.

But how did we arrive at the present state of knowledge about what chemicals *are*, as opposed to what they *do*? It has been a long and tortuous path. The nineteenth century was the key. It began with hazy and often inaccurate ideas about the material universe, and ended with the birth of triumphant new industries and intellectual pursuits based on the new science of chemistry. During that century, incorrect ideas about the nature of medicines, poisons, food, our own bodies, and all the other things that a lay person would call 'organic', some of them dating back to the ancients, were overturned.

There is a large gulf between what is intuitive about the material universe, and what has been discovered over the centuries about what is really going on. To take a mundane example: what happens when a new, shiny copper roof gradually turns first dark, then green? The answer cannot be discerned merely by consulting your senses, much as the modern mystics would like to be able to do so. When a modern chemist sees a sample of the brilliant red rock vermilion, he will tell you that it is made up of the shiny liquid metal mercury and the yellow powder sulphur. This seems extraordinary. The mercury and the sulphur are 'there', although the senses cannot perceive them.[4] How can this be true? What evidence is there for this remarkable assertion?

The substances with which we are surrounded have so many strange properties that early people inevitably associated them with different mythological spirits. The names cobalt and nickel derive from the German word *Kobald*, an imp, and from an old High German word for Satan ('Old Nick'). The early history of chemistry is inextricably bound up with magic and superstition, and was considered dangerous. In 1602 the *Accademia dei Segreti*, or academy of secrets, was compulsorily dissolved by Pope Pius III. The origin of the word *chemistry* is obscure, but one thing is certain: the word *alchemy* means exactly the same – it is nothing more or less than the Arabic *al-khemy*, or 'the chemistry'. The long period of what we now call alchemy, lasting through the Middle Ages and until the scientific renaissance of the seventeenth century, was ill-understood chemistry mixed up with mythology, religious belief and superstition. When gold dissolves in aqua regia (mixed nitric and hydrochloric acids), this was described in 1572 as a lion devouring the sun. When new initiates were instructed how to prepare an amalgam of mercury and lead, they had to do it at the moment of conjunction of the planets Saturn and Mercury, for it was thought that the metals known to the alchemists were under the influence of the planets. Such conceits lived on (and live on today) through the agency of sects like the Rosicrucians.

The alchemists had largely lost contact with a more productive line of speculation stretching back well before them. From very early times, philosophers considered that all matter must be made up of one or more fundamental ingredients, or elements. There were numerous variations on this theme. Anaximedes (500 BC) thought that everything was made of air, while Heracleitus, a little later, thought that everything was made of fire. At the other extreme was Anaxagorus, who thought that the number of elements was very large, or infinite; one for every different substance. In between (not chronologically, for they had lived centuries before), were the Chinese and Hindu philosophers, who each thought that there were five elements: in the case of the Chinese, water, fire, metal, earth and wood. The Greek Empedocles reduced this number to four. In what may have been the first chemical experiment, he proved that wood could not be an element because when it burned it turned into the true four elements, fire, ash (earth), air and water. Later on, the famous alchemist Paracelsus stated that everything was made of salt, sulphur and mercury.

Paracelsus had no evidence for this assertion. Experiments, not speculation, are the key to understanding. Even Aristotle, who has been criticized for never doing any actual experiments, said that 'Nothing is in the intellect that is not first in the senses', but this precept was largely lost sight of for many centuries. The universe is much more complex than people in the Middle Ages gave it credit for. They spent their time looking for instant solutions to the big questions, whereas true knowledge could come only from the painstaking accumulation of results over many years, starting with the simplest things.

The two goals of the alchemists, to turn base metals into gold and to achieve eternal life, were in fact one. They thought that wealth, health and longevity were united in the mystical substance known as 'the philosopher's stone' (which may not necessarily be a stone; it might be a metal, a red powder, or something quite indescribable). If only they could make this, all problems would be solved. The raw material for the stone was the Adamic or virgin earth, which occupied the lowest niche on the scale of perfectibility. Virgin earth was to be found everywhere, but only the adepts could transmute it. Human perfectibility, as contained in the teachings of Christ, was inseparable from the perfectibility of metals. Gold, the most noble metal, was immortal. 'Bring to me the six lepers, silver, mercury, copper, iron, lead and tin, that I may heal them,' said the alchemist Geber. Mercury, the only liquid metal, played a central role in the perfectibility of the metals and, through them, the human body and spirit, although most alchemists asserted that the true quicksilver of the adepts was not the same as common mercury or quicksilver. Human putrefaction after death could be reversed. Christ had done it, and so could the alchemists, if they tried hard enough. Perhaps distilling horse manure with egg yolks might be the solution?

No one could give a recipe for the philosopher's stone, though everyone knew of someone who had it, or rather had had it at some point. In the treasury at Vienna there was a large medal struck from 2½ lb of gold made from quicksilver at the court of the Emperor

Ferdinand III of Prague by the burgomaster, Count von Russ.[5] This with the aid of a single grain of red powder given him by a certain Richthausen, who had received it from an unknown source. The splendid Queen Christina of Sweden became very interested in alchemy, and wrote in 1667 of the successful transmutation of 500 lb of lead into gold by a Dutch peasant using another grain of red powder. She praised chemistry as 'a beautiful science, and the key which opens all treasures, giving health, glory and wisdom'.[6] She regretted that it had been degraded by charlatans, but said that it remained the royal science. The red powder of the alchemists had become inextricably enmeshed with her concept of her own queenly magnificence.

Charlatans there were aplenty. The diarist John Evelyn in December 1650 met in Paris 'An impostor that had like to have impos'd upon us a pretended secret of multiplying gold; 'Tis certain he had lived some time in Paris in extraordinarie splendour, but I found him to be an egregious cheate'.[7] Sixty years later, the Italian conman Count Ruggiero, after touring the capitals of Europe swindling various members of the nobility with similar schemes, was rumbled in Germany and hanged on a gilded scaffold.

Indisputable facts slowly emerged during the long period of alchemy, despite the obsession with these strange goals, and despite the best efforts of many of the alchemists to keep their discoveries secret. Many interesting new substances such as oil of vitriol (sulphuric acid) and saltpetre (potassium nitrate) were discovered. In the opinion of some, the false goal of the philosopher's stone was a help, not a hindrance. It would have been difficult for the alchemists to have kept up their enthusiasm without their mystical target.

By the time that Queen Christina made her pronouncement, modern chemistry was emerging. In that year Robert Boyle attained his fortieth birthday, and the Royal Society had already been founded for seven years. It was probably the invention of the printing press that made the greatest contribution to the transition from alchemy to scientific chemistry. The printed book demands verification.[8]

The final disappearance of alchemy took another century or so and is generally reckoned to be represented by the case of Johann Semler, Professor of Theology at Halle. In 1787 he purchased from a charlatan an elixir called *Luftsalz*, or atmospheric salt. Semler was told that, when it was kept warm and moist, gold grew on it, and this indeed seemed to be the case, for the metal was analysed by the Academy of Sciences in Berlin and found to be pure gold. Semler also sent the Academy a sample of the salt, which they found to consist of sodium sulphate (Glauber's salt), magnesium sulphate (Epsom salt) and urine. Undeterred, Semler wrote to them, 'Two glasses are bearing gold. Every five or six days I remove it; each time about twelve to fifteen grains. Two or three other glasses are in progress and the gold blooms out below.' A later sample sent for analysis, though, showed a disturbing loss of quality; what appeared gold now turned out to be pinchbeck. Semler's servant, who had been putting the gold in the glasses to keep the old man happy, was indisposed, and his wife thought that pinchbeck would serve just as well at a much lower cost.[9] Semler was clearly gullible. But even Joseph Priestley, the discoverer of oxygen, wrote to Benjamin Franklin to say that he 'Did not quite despair of the philosopher's stone', receiving the answer that, if he did find it, he should 'Take care to lose it again'.

So, by Semler's time, modern chemistry had already been struggling to emerge for a century and more. Angelo Sala (1575–1640) showed that, when a piece of iron is dipped into a solution of blue vitriol (copper sulphate) and becomes coated in copper, the copper comes from the blue vitriol; the iron has not undergone transubstantiation into copper.[10] New, more modern concepts about the nature of matter slowly began to emerge.

The emergence was slow. A few dozen more of the right experiments along Sala's lines would have explained so much, but people would keep doing experiments that did not prove anything in particular, or gave misleading results. Many scientists were still irresistibly drawn towards trying to explain the great, and complicated, phenomena; the weather, volcanoes, what the moon is made of. But this is to take far too much benefit from hindsight. The difficult part of progress lies in recognizing what are the key experiments, not in doing them; early chemists spreads their talents thinly.

The first great task of chemistry was to clarify exactly what was meant by the term element, and to say what were elements and what were not.

There are good reasons for calling Robert Boyle (1627–91) the father of chemistry; or, if Lavoisier was its father, then Boyle was its grandfather. He was the first person to *claim* to be a chemist. He said it was an 'art' that he would study as a practical philosopher rather than as a physician or an alchemist, and he said that he was going to set down the results of his experiments clearly.

In 1643 Torricelli had made the first vacuum, by inverting a tube containing mercury. The weight of the mercury creates a space at the top of the tube that is essentially empty; it contains nothing. Air and nothing therefore did not appear to be identical, although many did not accept this at first. The concept of gases was a novel one, and took some time to catch on, especially the idea that there could be more than one kind of gas, or 'air'. Boyle invented the vacuum pump and carried out numerous experiments with it, such as showing that a bladder of air weighs more in a vacuum than it does in air. These experiments firmly showed that air was a 'ponderable substance'; it was not a kind of universal nothing or 'ether'. Some people continued to struggle with the concept of a gas, though. Dr Johnson's refutation of Bishop Berkeley did not take into account that some matter is invisible and cannot be kicked. For the time being, air was also thought to be a single substance, and it took until the 1770s to disprove this.

The first important definition in modern chemistry came when Boyle, in 1661, threw out all earlier loose definitions of the term elements. He defined them as 'certain primitive and simple, or perfectly unmingled bodies . . . not being made of any other bodies, or of one another'. The practical test was whether a substance could be broken down into simpler substances. If it could not, it was an element. As to what were elements and what were not, Boyle had little to go on. Even until the late eighteenth century some people stuck with the Greek selection of earth, air, water and fire. Although Boyle had shown that air was what we now call a gas, this did not prove that it was not an element.

The ancients had thought that earth was produced when water was heated (as indeed it does to the eye when the water is hard). Could this explain the genesis of the hills and mountains, given enough time? Boyle indeed found that, when he repeatedly distilled pure rainwater in glass vessels, solid flecks kept appearing, although he thought that they came from the air that was able to penetrate pores in the glass.

The mysterious elemental fluid water had even more remarkable properties. The alchemist Van Helmont planted a 5 lb tree in a pot of 200 lb of carefully dried earth, and watered it for fifteen years, giving it nothing but pure water. At the end of this time, the tree weighed 169 lb and the weight of the soil was virtually unchanged. So not only rocks, but living trees too, were made of water!

It took another man to show the error of this, and to prove a great many other things. Antoine Lavoisier, born in Paris in 1743, consciously set out to spearhead a chemical revolution; that is, to make a revolution in chemical knowledge part of the revolution in society that was taking part throughout France. Ironically, this led eventually to his execution in 1794 as a member of the establishment and former collector of taxes.

In 1770 Lavoisier told of the results of his experiments on the theme of 'Whether water can be changed to earth as was thought by the old philosophers, and is still thought by some chemists of the day'. He showed that the particles of 'earth' that appeared in water when it was boiled did not come from the air, because the weight of the sealed flask plus water remained constant, and neither did they come from the water. But the flask got lighter, because the water was dissolving the glass.[11] (The fact that boiling water slowly dissolves soft glass is not obvious; it is only a slow process, but the early chemists had after all only been seeking an explanation of slow geological processes.)

Lavoisier turned his attention to the nature of air and fire, and what happens when you heat substances with or without air present. Experiments on heating things had been done since time immemorial; obtaining metals, for example, involved strongly heating their ores with other substances. The term 'calcination' (from the Latin word *calcis* for lime, obtained by strongly heating limestone) referred to any process involving very strong heating. Now limestone loses weight (in the form of carbon dioxide) when it is calcined, but coming as an extraordinary surprise to the alchemists was the observation that metals *increase* in weight when calcined (discovered in 1489, if not sooner). Their attempts to explain this are rather comical. An obvious explanation to them was that the metal contains something weighing less than nothing that escapes when it is heated. Paracelsus, for one, associated this with the concept of the metal dying during the calcination process, and called the residue remaining afterwards *terra damnata* and other pejorative terms.

The true explanation, that the metal combines with something (called a spirit by the alchemists), was also suggested at an early date. Boyle thought that when the metal was calcined it was absorbing 'corpuscles', but surprisingly, in view of his experiments on the nature of air, he did not specifically state that this was where the corpuscles came from. Stephen Hales in 1727 showed that when lead was calcined in air it gained weight when it formed the red oxide (or calx, as it was then called), and that, when the calx was reheated, the original amount of lead was regained. But he did not study the gas given off, assuming that it was air.

Unfortunately, for the time being these findings were lost sight of and chemistry took a detour with a new version of the first hypothesis, that burning bodies give something off; they do not absorb it. Phlogiston was the name given to this essence of fire or inflammable principle. A metal was calx plus phlogiston, and when it was calcined the phlogiston was driven off leaving behind the pure calx or earth. Gold could not be calcined because the phlogiston was too tightly held. Phlogiston held sway, despite the difficulties already raised by the weighing experiments. One of phlogiston's strongest advocates, G.E. Stahl, was initially able to get away with flatly denying that the calx of a metal weighed more than the metal, because that contradicted the theory. We shall meet some later chemists who hung on desperately to their theories when all the facts were against them. Rather than abandon his theory, Stahl came up with the alternative explanation that phlogiston weighed less than nothing. Newton had by now said that everything in the universe attracts everything else and that for something to weigh less than nothing was impossible, but perhaps he was wrong; perhaps there could indeed be a substance on which gravity acted in the converse sense. Perhaps some other property such as colour might be more important than weight.

It eventually became clear that Newton was right and that there was no such thing as negative weight. Weight was the most important thing to measure before and after a chemical change. Hard-headed commerce might have given more clues. Around 1700 the apothecaries were selling cups made of antimony – an element of great interest to the alchemists because it could be used for purifying gold,[12] and therefore had its part to play in the ladder of purification of metals, and of human perfectibility. If you left wine in an antimony beaker overnight, it would turn into a most satisfying (and dangerous) purgative. Some of those selling the cups claimed that they would last forever, never becoming thinner with use. Vigani, apothecary and professor of chemistry, said that this was nonsense. Why? Not because he had proved that the wine, after standing overnight, contained antimony that could only have come from the cup. Not because he had weighed a cup before and after use, and showed that it had got lighter. But because he scornfully pointed out that, if the antimony was never used up, 1 oz of it ought to last every apothecary for ever, so why did they keep buying it?[13]

The prevalence of the phlogiston theory throughout the eighteenth century did not inhibit experimentation; like the incorrect theories of the alchemists, it encouraged it. Important discoveries were made during the phlogiston period, including new gases such as hydrogen (initially called 'phlogistated air') and oxygen. But theory lagged behind, and no one knew quite what they were doing. By the end of the century so many contradictions had accumulated that the phlogiston theory began to collapse under its own weight. Heating mercury calx (mercuric oxide) in a sealed flask gave mercury; where had the phlogiston come from? This had to be explained by saying that mercury calx was not a real calx, but something that looked like a calx, and so on. The level-headed Joseph Black (1728–99) said with complete justification, 'Chemistry is not yet a science. We are very far from the knowledge of first principles.' Chemistry during the phlogiston era is a fascinating study for historians, but it is difficult to avoid the recurring feeling that it was a waste of most of the eighteenth century.

The great achievements of Lavoisier lay not so much in the experiments that he did (and he tried to steal the credit for discovering oxygen from Priestley and Scheele), but in his demolition of the phlogiston theory and the establishment of the conceptual framework of modern chemistry.

He worked out the nature of the first of the four ancient elements, air. When he heated a retort full of mercury for several days over a hot flame, it became covered with red flecks of the calx or oxide. He connected the arm of the retort to a reservoir of air and could thus measure the air taken up. However much mercury remained, one-fifth of the air was used up and no more. The remaining gas extinguished a candle and suffocated a mouse. Thus air was a mixture of two components, one of which was the sluggish gas that he called 'azote' (still its name in French), and that is today called nitrogen. He collected the mercury calx and, by heating it strongly, obtained the other component, a gas that made a candle flame burn brilliantly and that he named 'oxygen'. Oxygen, nitrogen and mercury were assumed to be elements. The calx is an oxide of the mercury. Phlogiston was abandoned; Lavoisier had no truck with anything that weighed less than nothing; phlogiston was 'the spirit of minus oxygen' and got the boot.

Lavoisier defined an element along Boyle's lines, but stated it more clearly. In 1789 he published a remarkably modern looking list of elements. It contained a few errors; in particular, he continued to include heat, which he called 'caloric'; the true nature of heat continued to baffle physicists until well into the nineteenth century. He thought that lime and magnesia and baryta were elements, whereas they are the oxides of the then unknown elements calcium, magnesium and barium, which were obtained as metals by Davy fourteen years after Lavoisier's death. A rather more serious error was his assumption that all acids contain oxygen; and that therefore hydrochloric acid, or 'muriatic acid', must do also. Chlorine, discovered by Scheele in 1774, therefore could not be an element but must be 'oxymuriatic acid', a compound of oxygen. This assumption caused much confusion until corrected by Davy. But Lavoisier's approach stands out in contrast to that of the alchemists, or the phlogistonists. Instead of *asserting* that lime and magnesia were elements and attempting to shoehorn a mass of inconvenient facts to accommodate this, he stated that every element on his list should be considered such until someone was able to decompose it. Not a big step on from Boyle a century before, but presented with startling clarity. In a statement that presaged the discovery of the subatomic particles more than a century later, he said that the elements might themselves be compounded of simpler substances, but they were *entitled to be considered* as elemental while no means existed to break them down. His list ran to thirty-three elements, of which twenty-three, including the then recently discovered tungsten and molybdenum, have stood the test of time.

The second ancient element, earth, faded away when it became clear that there were various different 'earths' having distinct properties; for example, mercury oxide might be considered an earth according to the ancient definition. Calcination of various other earths soon led to the discovery of further new elements such as niobium (1801).

The composition of a third primeval element, water, was clearly demonstrated by Henry Cavendish in 1781, who showed that water and nothing else was produced when hydrogen (which he thought might be identical with phlogiston) and oxygen ('dephlogistated air') were burnt together. Lavoisier repeated the experiment on a large scale and showed that 6.6 parts by weight of oxygen combines with one of hydrogen. This was an experimental error; the correct figure is 7.94, or, to the nearest whole number, 8; an important ratio. Lavoisier correctly stated that water is a compound of two of his elements, hydrogen and oxygen. Soon afterwards, the highly eccentric Cavendish, a brilliant experimenter, was accepting that Lavoisier's new explanations not involving phlogiston might well be true. By the end of the century, Lavoisier's team had won the match. To celebrate the victory, the plutocratic Lavoisiers had their portrait painted by David. An allegorical play was performed in which the deformed figure of Phlogiston, weakly defended by Stahl, was put on trial by the handsome Oxygen, found guilty and burnt at the stake, when, of course, he disappeared, since combustion of phlogiston logically leaves nothing.[14]

Lavoisier's was not a particularly attractive personality. Most people have found it easier to admire his attainments than to like the man. There are similarities between his behaviour and that of another, August Kekulé, who came to prominence several decades later.

Antoine Laurent Lavoisier was the son of a Paris lawyer and became a member of the French establishment. He combined his scientific researches with climbing the greasy pole of pre-revolutionary French society. Although chemistry was his consuming passion, he was a man of affairs, qualified with a law degree and taking an active part in politics. He had no children when executed at the age of fifty, and appears to have had few personal friends. His wife, Marie-Anne Paulze, only thirteen at the time of their marriage, was the daughter of a Fermier Général (territorial administrator and tax collector). Her family connections contributed to her husband's rising to a similar post in 1780. This made their fortune, but when after the Revolution he and the other Fermiers were tried on trumped-up charges such as adulterating snuff and hazarding the health of the citizens, her Marie-Antoinette-like arrogance may have sent him to the guillotine.

Although this rather cold man was apparently scrupulous in his personal dealings, he has been much criticized for the way in which, with his legal training, he shaded the record of his researches to allow extra credit to drift his way. The main journal in which he published, the *Mémoires of the Académie des Sciences*, was published up to four years late, so in order to establish priority for his discoveries he quite reasonably published a preliminary report in a faster journal, just as many scientists do today. But then, when he wrote the full report, he often implied that the original work had contained things not known to him at the time. Then he got another bite at the cherry by revising again when the experiments were published in a collected format. He used a lawyer's turn of phrase to finesse the truth. For example, he would write, 'I published this in the *Mémoires* of the Academy, 1781' when the *Mémoires* for 1781 had not been published until 1784. In 1776 he said, in summary, 'Oxygen is what I obtained from the red oxide of mercury, and which M. Prisley [*sic*] obtained from a great number of substances by treatment with spirit of nitre.' This implied that he had obtained it first, whereas Priestley had made it in August 1774 and told Lavoisier about it two months later. Another technique used by Lavoisier was to write an account of the contributions of earlier workers as if it were a dispassionate record, but to exaggerate the status of the second-rate (the alchemist and tree-grower Van Helmont), while abbreviating and misreporting the work of his true competitors (Cavendish, Black). There are real similarities here to what Kekulé's critics claim that he did more than a century later.

The period following Lavoisier's untimely disappearance from the scene was marked by the final burial of the phlogiston theory and by important further discoveries. Chemistry as it was generally understood around 1800 meant inorganic chemistry: metals, their oxides and other compounds, the various gases, water, minerals, new elements. This chemistry was on the right track and its findings gradually began to permeate technology and commerce.

This was inorganic chemistry, the chemistry of the stone. But there was another kind of chemistry, the chemistry of the leaf, that would eventually have an even more profound effect on the world. At number 8 on Lavoisier's list of elements was the unglamorous substance 'charbone', or carbon, much less exciting in its elemental state than most of the others. Stahl in 1703 had thought that charcoal was 'a sulphur composed of phlogiston and aerial acid', but Lavoisier thought it was an element and put it on his list. It was a 'sleeper' that was later to generate whole disciplines, whole libraries of discovery and thought, all to itself.

TWO

Sugar is Nothing but Water Rendered Solid

Mr Gresham, with all his faults, is a gentleman; and how he can talk about his affairs with a low apothecary like that, I, for one, cannot imagine.

Anthony Trollope, *Doctor Thorne* (1858; Oxford edn 1981)

During the long centuries in which the alchemists, the assayers and other tinkerers with metals and stones gradually built up their half-knowledge of the inorganic world, other vitally important gifts of nature slowly emerged from the shadow. These were the natural drugs with which to kill or cure. In reality, the two strands of knowledge were part of the same rope, but there was no true understanding of the way in which they were related until the second half of the nineteenth century.

Drugs, like alchemy, were shrouded in secrecy. If there was one area of primitive science offering even more power and influence than the ability to smelt gold, then surely it was familiarity with those feared substances that might cure the King's sickness, or dispose of his enemies.

Throughout its infancy, chemistry was largely the handmaiden of medicine. Modern chemists rightly emphasize the achievements of Boyle, Lavoisier, Priestley and the other 'natural philosophers', because they made the discoveries later seen as important keys to understanding. But the vast majority of those who dabbled in chemical experiments were apothecaries or connected in some way with the medical profession. There were many pharmaceutical tradesmen, and all gradations in between. Some who sold medicines during the day dabbled in chemical experiments at night.

Physicians and apothecaries, like the alchemists, tried to relate everything to everything else, and there was a great deal of dispute about different systems of medicine. A central figure in this dispute was Paracelcus (Theophtrastus Bombastus von Hohenheim), born in Swabia in 1493 and from whom we derive the word 'bombastic'. Until he arrived on the scene, most doctors administered herbal medicines, in a tradition going back at least to Galen, physician to the Roman Marcus Aurelius. Medicine was for a long time dominated by Galen's four humours: blood, phlegm, yellow bile and black bile. Each humour was associated not only with a particular temperament (sanguine, phlegmatic, choleric or melancholic), and with a particular organ of the body, but also with one of the four elements and even with one of the seasons of the year. Galen taught that disease is caused by imbalance in his humours, and can be fought by administering drugs having the opposite qualities, such as a drug with cooling properties to treat fevers. Galenists held that nature produced a plant to cure every malady, a view that persists today. Many plants were ascribed medicinal properties on no evidence other than their appearance. Liverworts, shaped like tiny livers, were therefore administered for presumed diseases of this organ and hence linked through occult medicine and alchemy to universal phenomena such as the rising of the stars. The fact that liverworts have virtually no real medicinal value was less important to the physicians at first than was the maintenance of their particular belief system.

Paracelcus was either an innovative scientist or some kind of nut, or more probably both. He was the first to recognize that drugs were really chemicals, just as metals are. He said that the methods of the alchemists should be applied to discover new medicines, and that these should replace the herbal medicines. He recognized that the human body itself functioned by chemistry, although the way that he described this happening was bizarre; the body contained the Archaeus, or internal chemist, sometimes illustrated as a kind of homunculus with arms and head but no legs, that carried out distillations on the components of food. This was not really so bad a picture, it just reflected the fact that virtually the only chemical operation that Paracelsus knew about was heating things. He also correctly said that there is no real distinction between foods, drugs and poisons, and that one could be poisoned by the food that one ate. They are all chemicals, really. 'It's all a matter of the dose,' as Paracelsus correctly maintained.[1]

The Paracelsan system, called Iatrochemistry, did not completely replace herbal medicine, and from then onwards they existed side by side, with frequent vitriolic disputes between their practitioners. Other thinkers were more conciliatory; after Boyle came up with his idea of an element, there were various attempts to reconcile Galen's system with chemistry. Some maintained that phlegm and faeces were elements, additional to or replacing the traditional Greek four elements and thus bringing the chemical elements under the same roof as the four humours.

Most herbal medicines were elixirs: complicated concoctions taking many hours to prepare and correspondingly expensive. These were known as 'galenicals', as opposed to 'simples', which were the extracts of a single herb. In England, the best-known physician (as well as astrologer, iconoclast and general nuisance) of a slightly later date was Nicholas Culpeper (1616–54). His *Complete Herbal and English Physician* was the most popular compendium of herbal medicine for many years. Culpeper's 1653 edition lists over 300 medicinal plants used in his medicines, and his 'Venice Treacle' is a concoction of no fewer than sixty-nine different herbs and venoms.

The majority of people who handled the natural products were the apothecaries, who battled for recognition as a true profession. The trade of apothecary was always a peculiar one. Perched uneasily in the no man's land between the senior professions such as physician and the normal run of tradespeople, they had a large amount of empirical and sometimes dangerous knowledge. They lacked the status of the physicians, but had at least as much power. The attitude of the establishment towards the apothecaries in say the eighteenth century reminds one of that between the directors of an ill-run modern corporation and the scientists in its research department. They were feared and mistrusted, kept in their place, essential to the functioning of the enterprise but with no one capable of telling whether they were doing a good job or not.

Everyone wanted a slice of the action in the valuable trade in drugs from abroad. In England, the apothecaries were for many years denied their own royal charter and kept under the thumb of the wealthy Grocers' company, which was the importer of drugs and spices from east and west. The grocers had an official called the Garbler of Spices who was allowed to enter any shop or warehouse to garble – that is, check the purity – of spices.[2] They used this as a device to keep the apothecaries in line, although the apothecaries could complain, with complete justification, that, when it came to checking the purity of drugs, the grocers did not know what they were doing. (Neither, really, did the apothecaries.) When they got their own charter in 1614, the apothecaries immediately began to get their own back on the grocers by dealing in Malmsey, sack and bastard under the pretence that these were medicinal remedies. They may have made a tactical blunder when fighting for their independence from the grocers, for they had enlisted the support of the physicians, who now started to treat them as their assistants. The succeeding two centuries and more were marked by a continuation of this jostling for power, each group trying to exclude others from handling the valuable drugs.[3] It was important for the apothecaries to prove their credibility and protect their professional status by persecuting those who sold inferior wares. One Eason, a frequent offender, was fined £6 13*s* 4*d* in 1616 for selling his 'Mithridatie' and 'London Treacle', a fine that was inexplicably reduced on appeal to fourpence.[4] What he said in his defence is not recorded, but it is tempting to imagine him saying, 'You say that the rubbish I sell is worse than the rubbish you sell. Prove it.'

Charles Darwin's grandfather, Erasmus Darwin, in a letter to his son dated 1790, gave a later account of the style of an incompetent provincial apothecary:

> I remember a very foolish, garrulous apothecary at Cannock who had great business without any knowledge or even art, except that he persuaded people he kept good drugs; and this he accomplished by only one stratagem, and that was by *boring* every person who was so unfortunate as to step into his shop with the goodness of his drugs: 'Here's a fine piece of assafoetida, smell of this valerian, taste this album graecum. Dr Fungus says he never saw such a piece in his life.'[5]

This went to the heart of the matter. Who was to say the man was wrong? The apothecary's reputation was based on a mixture of empirical knowledge, more or less folklore, and marketing skills.

The man's marketing strategy was actually quite a good one. People were bored stiff, but believed him.

On the continent of Europe, the status of apothecaries seems to have been slightly higher. Spacious laboratories were constructed in the seventeenth century for extracting and compounding pharmaceuticals, but invariably associated with a shop from which the public could buy ointments, purgatives and lead powders to use as cosmetics. There were cabinets of barks, seeds and dried leaves from all over the known world, together with the other items considered of great pharmaceutical importance. The largest of these cabinets became veritable museums; in 1660 the botanist John Ray visited Verona, where he saw a great museum of pharmacy that survived for another century before disappearing.[6] But, just as in England, the apothecary remained a figure of suspicion. Engravings by Teniers and many others usually show him, not as a serious investigator, but as a figure somewhere between ghoul and buffoon, surrounded by an array of macabre specimens – snakes, frogs, insects and the inevitable skull. Such engravings had a moral: the message to the public was about the unwisdom of giving large amounts of money to those who claimed to be able to cure, but whose craft was largely mumbo-jumbo.

The laboratories inevitably centred on the furnace, which remained the most important item of equipment for all chemical operations. In England, Isaac Newton spent a considerable amount of time perfecting his own designs for furnaces, which he could build without the aid of a bricklayer. His researches were largely what we would now call alchemical. The University of Cambridge tolerated them, just, but it could not be seen openly to support activities that were aimed at obtaining divine knowledge. This would cut out of the loop the clerics who ran the university and were God's official mouthpiece. When Giovanni Francesco Vigani was appointed to the new chair of chemistry at Cambridge in 1702, he took over Newton's laboratory. He carried out some researches, and he gave lectures instructing medical students in the practical art of dispensing: but he also ran a business as apothecary based at Cambridge and Newark.[7] He used his furnaces to prepare his most profitable remedy, a mercurial ointment to cure syphilis. His university post was unpaid and his income was made from teaching and by trading.

Campbell, in his highly entertaining survey, not long afterwards, takes us on a walkabout through the streets and shops of London. He pauses to describe the dispute between the two trades of chemist and apothecary:

The Galenists, by compounding the gross simples [crude drugs], are at continual War with the Chymists; they alledge, and perhaps not without Reason, that the Fire alters the Nature of the Subject, and bestows on it Qualities that did not before exist; that most Chymical Preparations were of such a fiery corrosive Nature, that they inflamed the Bowels, and set the whole System in Combustion; that Nature had provided simple Remedies for all Diseases, and consequently there was no Reason to have recourse to Fire. . . . The Chymists, on the other hand, alledge that to effect a cure by the Galenic Practice required such enormous nauseous doses, that Nature was oppressed. . . . Distillation and Sublimation must be perfomed in the Stomach before any Cure can be performed, and that consequently, it was saving more than half the Strength to perform these Operations beforehand. The Galenists are daily losing ground . . .[8]

There is a grain of truth in both sets of assertions. Most drugs in use today are not inorganics. The plant world would in the long run provide a far richer arsenal of drugs than the world of the alchemist with its lead, mercury and red-hot retorts. On the other hand, the Galenists were wrong in their belief that nature provided a cure for every ailment without human intervention. Crude plant extracts might have exceedingly useful medicinal properties, but there was no law of nature that says this has to be so.

During the seventeenth and eighteenth centuries, anatomy and surgery developed apace. The structure, although not the function, of the internal organs became well known. Surgeons could cure some diseases, if the patient did not die of shock or infection. The physicians, though, were far behind. The best they could do was summed up by E.R. Arnaud in 1650: 'In the most obscure, difficult and dangerous maladies it is better to hazard a doubtful remedy than to give nothing at all.'[9] They had some vague knowledge of what would happen if a particular organ was defective, but no knowledge of how it worked. To get this, they needed two new sciences. The first was physiology, the proper scientific study of what those organs actually did, and the second was chemistry.

Some physicians rejected chemistry, and in fact virtually the whole of what we would now call science, altogether. One of these was Philippe Hecquet. According to him, the body was a machine. It functioned through the activity of moving parts having nothing to do with chemistry whatever. 'Chemists write of bizarre fermentations, while we discuss disordered fibres,' he wrote in 1712: 'In these systems of the bile, the saliva, the blood and melancholy, of acid and alkali, volatility, fixity, aqueous, sulphureous, spirituous and phlegmatic, all that is really needed is a knowledge of solid and liquid, and in the place of all the chemical faculties, qualities and flavours we need only resistance to forces . . . we insist that all these qualifications are imaginary.'[10] Even in the nineteenth century, chemistry was held more or less in contempt by most medical men. The French physician Armand Trousseau (1801–67) deplored the 'vanity and pretentions of the chemists, who believe they can explain the laws of life', and advised those taking up medicine not to waste too much time on it. 'The acts . . . of animal life are subject to laws which ought to be regarded as essentially different from those which govern inorganic matter . . . in living organisms chemistry is controlled by special powers, which give it special direction'.[11]

There seems to be something about chemistry that generated opposition among some people; if not to the whole subject, then to large chunks of it; atoms, molecules, what have you. We shall meet fresh generations of nay-sayers in the guise of the antiatomists of the nineteenth century.

Chemistry would eventually teach the sceptical physicians that time spent studying it was not wasted, and that the drugs that they handled every day could not be properly understood without its help. This process was a long one.

The disappearance of phlogiston at the end of the eighteenth century led to a surge in discoveries, and chemistry became an immensely popular pastime. In England this was centred on the newly founded Royal Institution just off Piccadilly in central London.

The Institution was founded by Benjamin Thompson, a royalist American born in Massachusetts and a friend of Benjamin Franklin. He was twenty-three at the time of the American revolution, when he spied for the English and had to flee the country. In Europe he settled for a time in Bavaria, and was created a Count of the Holy Roman Empire by the Elector Palatine. Count Rumford, as he now became, was a highly practical and accomplished scientist who did much to clarify the puzzling nature of heat by establishing that it was a kind of motion, and also made practical inventions such as improved stoves and chimneys.

In Munich, he founded something typical of the age of enlightenment, an establishment for both employing and educating the poor by teaching them to apply the new practical discoveries. When he moved to England in 1795, he proposed something similar, and the Royal Institution was opened by public subscription four years later. Rumford himself did not stay long; he fell out with the management and left for Paris, where he married Lavoisier's widow. The marriage was not a success. Despite his adventurous past, Rumford was apparently a bore who expected to be kept in comfort on his wife's money, and within a few months he was locking her friends out of the house, and she was retaliating by pouring boiling water on his plants.[12]

As far as chemistry was concerned, the Royal Institution got off to a poor start. The first professor was Thomas Garnett, a pupil of Black's from Glasgow. This seemed to be a good choice, for the Glasgow tradition of chemistry was extremely strong and Garnett had a reputation as an excellent lecturer. But his wife had died in childbirth shortly before he came to London, and Garnett, labouring under depression, was not a success. Only three years after taking up the post, in 1802, he died of typhus contracted while working in a dispensary.

The new appointee was Humphry Davy, a Cornishman who had come to London as Garnett's assistant without any kind of degree and who was still only twenty-four years old.[13] He rapidly became an immensely glamorous and popular figure, chemistry's Byron. His poetry was praised by Southey and Coleridge and he was a charismatic lecturer. When at one point he fell ill, the crush of carriages in Albemarle Street bearing young female Davy groupies became so great that the staff of the Institution were obliged to erect a blackboard outside carrying up-to-date medical bulletins and temperature charts.

Before coming to London, in 1799, Davy had discovered the effect of breathing the gas nitrous oxide ('dephlogistated nitrous air'), and showing the effect of this 'laughing gas' on volunteers became his most famous demonstration. This clearly took chemistry into the realm of medicine. Davy aimed his well-prepared lectures at the intelligent lay person, who also needed to be well-off, for tickets cost the enormous sum of £20 per head. Only some were there to be entertained, and only some because of a burning interest in pure science. Many must have been there because they sensed that the chemists were on the brink of momentous discoveries about disease and the human condition. After all, if the absurd notion of a gas that made people light-headed and hysterical could be demonstrated to public satisfaction, what other extraordinary medicinal feats might be just around the corner?

Samuel Taylor Coleridge was one of those who described the effects of the gas:

> The first time I inspired the nitrous oxide, I felt a highly pleasurable sensation of warmth over my whole frame, resembling that which I remember once to have experienced after returning from a walk in the snow into a warm room. The only motion which I felt inclined to make, was that of laughing at those who were looking at me. My eyes felt distended, and towards the last, my heart beat as if it were leaping up and down. On removing the mouth-piece, the whole sensation went off almost instantly.[14]

Despite this early march of chemistry, though, scientists remained totally at a loss when faced with the extraordinary abilities of the plant world, which put all their efforts to shame. Edme Mariotte's *Essai de la végétation des plantes* (1676) was an early attempt to bring current chemical theories to bear on what plants were made of and what they could do. According to him, plants were composed of 'principles grossiers et visibles' such as water, sulphur, oil, salt, saltpetre, ammonia and earths. These 'principles grossiers' are themselves composed of simpler elements; for example, saltpetre is

made up of its phlegm or insipid water, its spirit and its fixed salt –
virtually meaningless concepts. Any attempt truly to understand
what billions of plants routinely did every day, to grow, reproduce
and produce drugs, was still far away: 'Any kind of plant can be
grown in a pot of earth by the action of rainwater and the principles
of the different plants are therefore the same . . . since the substances
separated by the distillation of different kinds of plants are the same,
chemistry can give no explanation of the differences.'[15]

What he says is partly true. Various French academicians, starting
from shortly before the date of Mariotte's assertion, had investigated
the effect of distilling numerous plants using fierce heat. Every plant
they tried gave the same result, so that by 1695 another chemist,
Wilhelm Homberg, could already state that the experiments were
worthless. This was true, but Mariotte was wrong to infer from this
that no explanation could be given by chemistry. A different kind of
chemistry was needed, but it would take 100 years more for it to
emerge.

One of the concepts that was needed was a better idea of what was
meant by 'purity'. If I were to tell you that ordinary white sugar
from your kitchen shelf is a pure organic compound in a very high
degree of purity, whereas shampoo or olive oil, for example, are
complex mixtures, you would probably accept my statement. Take it
as read that I have no particular reason for telling you an untruth,
and consider for a moment why you believed me. Underlying your
acceptance is your fairly sophisticated modern appreciation of what
constitutes a pure substance, based in turn on the knowledge that,
although you may not understand what is involved in defining one,
there are people in white coats around the world who do. Shampoo
you know to be a product produced in a factory somewhere by
mixing together the various ingredients shown on the bottle. The
case of olive oil is less easy. Here you are implicitly relying on the
expertise of analytical chemists who tell us that it is a mixture. They
would say that, whereas in a bag of sugar all the molecules are
identical, in olive oil they are not; they have various instruments that
tell them so, and various machines that, if necessary, could separate
olive oil into its components. As a modern person you are implicitly
comfortable with the concept of 'molecule' even if you may not

realize it. You are, for example, used to reading countless news reports of contaminants in the environment, and are aware that certain molecules have turned up where they ought not to be.

Sugar is what a modern chemist would call a 'natural product' in the strict, chemical, sense of the term. It is a single chemically pure substance produced by various plants. Olive oil, although a 'natural product' in the wider sense as understood by a dietician or a chef, is a mixture. There was no way for the early chemists to distinguish between these two degrees of 'natural product' until they had developed better ideas of what constituted chemical purity.

While we are on the subject, let us get one other thing out of the way. When a chemist makes sugar in the laboratory, there is absolutely no difference in its properties, physical, nutritional or toxic, from the sugar in your kitchen, providing it is pure enough. No one makes sugar (sucrose) synthetically, because it would be ludicrously expensive compared with what the sugar cane or sugar beet can make. But the fact remains. And brown sugar is white sugar plus various kinds of impurities, what a chemist, if he wanted to be provocative, would call dirt.

The term 'organic chemistry', meaning the chemistry of the products of living organisms, crept into use from about 1777. As the new nineteenth century dawned, its study was at a very rudimentary stage compared even with the modest state of knowledge that inorganic, or mineral, chemistry had attained, and had hardly advanced since Mariotte's day. Chemistry texts of the period do little more than divide organic chemistry into vegetable and animal chemistry, and list the various 'proximate principles' that constituted their domain: gum, saliva, urine, fibrin, albumen, gelatine and so on. This was a heterogeneous ragbag of no use at all. Some of the substances on the list, such as indigo, were what we would now recognize as pure organic compounds (more or less); others, such as blood, were impossibly complex living systems, or, like 'bitter principle', were so ill defined that they could encompass a vast repertoire of possible substances.

Once more, it was the innovative Lavoisier who can be credited for carrying out some of the earliest experiments of real worth. Initially he followed Mariotte and the others in doing some distillation experiments on organic materials, but, although he stated the issues much more clearly than Mariotte, for the time being he was as mystified as the rest. 'We still do not know', he wrote in his laboratory notebook, after trying the effect of furnace heat on some oak wood, '1) What is the quality of that immense amount of air that is released during distillation . . . 2) The nature of oil; it appears that it can be reduced to air and water by combustion, but we know nothing more; 3) what charcoal is . . .'.[16]

From about 1785 he began to carry out some more sophisticated analyses of organic materials, such as spirits of wine (impure alcohol) and olive oil, in order to find out which of his list of elements they contained. It must have come as a considerable surprise to him to find that not only were the natural products composed of the same elements that were found in the mineral world, but that the range of elements found in them was severely limited. Various of his analyses convinced Lavoisier that plant substances such as sugars were composed of water together with carbon, or what he initially called *matière charbonneuse*, and *nothing else*. Their mysterious nature seemed intimately associated with the properties of the dull substance carbon, best known as charcoal, ubiquitous and known since ancient times, and seemingly so mundane.

In the analysis of sugar that I made I have recognised that this substance, and all those analogous to it, are composed of nothing but water and charbon. The small quantity of other principles ['such as earth' is added in the margin] are not essential to sugar; they do not form a constituent part. Sugar is therefore nothing but water rendered solid by combination with charbon . . . one of the principal operations of vegetation is to join the carbon to the water . . . one ought to regard plant substances as a combination of oxygen, hydrogen and charbon, a combination of little solidity that a very slight heat can alter and destroy. . . . One must know that there can be no vegetation without water and without fixed air, or *acide charbonneux* [carbon dioxide]. These two substances mutually decompose during the act of vegetation . . . the inflammable principle separates from the oxygen to unite with the charbon in order to form oils, resins and to constitute the plant.

At the same time the oxygen which becomes free combines with light to form vital air . . .[17]

He does not describe in detail the experiments that led him to this very accurate description of photosynthesis and the manufacture of natural products by the plants. In modern terminology, we would say that the plant takes in carbon dioxide from the air and water from the soil, and using the energy from sunlight converts them into its own tissues and into all the natural products that it produces. In doing this, the plants provide all the nourishment that animals, including man, require. At first it was thought that there were major differences in the elemental requirements of plants and animals and that, in particular, animals required nitrogen and plants do not. This was later disproved. Both plants and animals contain smaller amounts of other elements than carbon, hydrogen and oxygen, and many natural products contain these extra elements, including, for example, nitrogen. But the fact remains that important natural products such as alcohol, sugars, acetic acid (from vinegar) and many others do not. They contain only the three elements, carbon, hydrogen and oxygen.

As Liebig later said, man, the being who performs all of the wonders of thought, is formed of condensed air; he lives on condensed and uncondensed air, and clothes himself in condensed air.

But the strangest part of the matter is, that thousands of these tabernacles formed of condensed air . . . destroy each other in pitched battles by means of condensed air . . . many believe the peculiar powers of the bodiless, conscious, thinking, and sensitive being, housed in this tabernacle, to be the result, simply, of its internal structure and the arrangement of its particles or atoms; while chemistry supplies the clearest proof that . . . man is, to all appearance, identical with the ox, or with the animal lowest in the scale of creation.[18]

Yet we have still not reached the limits of simplicity. Other natural products do not even contain any oxygen; they are hydrocarbons, containing just two elements. To most people the term 'hydrocarbon' means petroleum and natural gas, but naturally occurring hydrocarbons also include delightful substances such as turpentine and lemon oil, both mixtures of fragrant hydrocarbons. Explaining all this was going to be a major undertaking. Organic chemistry seemed to be at the same time astoundingly simple and monumentally complex.

Carbon seemed to hold the key. It is rather intractable in its elementary form, the black powder of charcoal or lampblack. When heated in the absence of air, it refuses to melt at any temperature accessible to the early chemists. It is opaque, and insoluble in all solvents. It does not have a crystalline structure like most chemically pure solids, but the atoms are arranged randomly, as they are in glass, another example of an amorphous material. There were strict limits to the useful experiments that Lavoisier and his contemporaries could carry out on a substance that they could not melt, dissolve or see through.

Carbon does, however, have a crystalline form, although it is much rarer than the material that is all around us in a greater or lesser degree of purity as soot, coal or charcoal. In 1797 Smithson Tennant, the eccentric discoverer of the elements osmium and iridium, sealed a weighed quantity of diamonds in a gold tube with air, and heated them strongly. He obtained 'fixed air' (carbon dioxide) and nothing else, and the quantity of fixed air was exactly the same as that obtained from an equal weight of charcoal. Therefore, diamond 'consists entirely of charcoal, differing . . . only by its crystallised form'. This experiment, although remarkable, went no distance towards resolving the mystery of carbon. If anything, it only deepened it. Diamond is what we would now call an allotrope of carbon, that is a form of the same element differing in some way in arrangement of the atoms – if you believed in atoms, that is.

The number of true natural products known at the time Davy gave his lectures was very limited. Benzoic acid from gum benzoin and one or two others had been known for a long time. But many more true natural products were out there to be found, if you looked for them in the right way.

THREE

Try Some of these Coffee Grounds

Holmes is a little too scientific for my tastes – it approaches to cold-bloodedness. I could imagine his giving a friend a little pinch of the latest vegetable alkaloid, not out of malevolence, you understand, but simply out of a spirit of enquiry in order to have an accurate idea of the effects.

Sir Arthur Conan Doyle, *A Study in Scarlet* (1887)

Since ancient times, the sea routes between Arabia and India, across the Arabian Sea, had been busy with the spice trade. Until about the seventeenth century, there was hardly any distinction between drugs, herbs and spices. The ancient herbal medicine systems, for example, that of the Indian surgeon Susruta the elder, in the fourth century BC, recommended what we would now call spices, such as ginger and mustard, to cure various ailments. So the trade routes by which drugs and spices eventually reached the West were identical. In ancient times this was either overland through Egypt or by water through the Red Sea, which was controlled by the Arabs, and so on to Venice and Salerno.[1] Susruta's view that there is no real distinction between foods, drugs and poisons is scientifically correct, and the same conclusion was reached in the West by Paracelsus.

Later the Portuguese opened up the sea routes around Africa. The cost of this valuable trade in human life was enormous. Between 1412 and 1640, 956 Portuguese vessels sailed to Africa and India, of which approximately 150 were lost, carrying at least 100,000 men.[2] Trade with India was later controlled by the East India Company, which at one point employed no fewer than twelve million people in producing and collecting goods for Western markets. The early fortunes of the Company were intimately bound up with medicine and the trade in drugs, and the medical expertise of the British doctors was eagerly sought by the Indian rulers. Gabriel Boughton, a surgeon in the Company's ship *Hopewell*, transferred to the court of the Emperor Shahjahan at Agra from 1645 to 1650, and a little later, in 1716, Company surgeon William Hamilton cured the Mogul Emperor Farrukhsiyar of a painful disease, which had delayed his marriage.[3] The Company was rewarded with valuable trading permits, which gave it an advantage over the local traders.

The export of high-value low-volume goods such as drugs (known as 'fine goods') was a welcome diversion for the Company away from 'gross goods' such as textiles and the cheaper spices, and could substantially increase the value of a cargo. The trade routes, though, were hazardous because of piracy, acts of war and shipwreck. In the 1690s the French captured 4,200 English merchant ships in eight years; and during its early days the East India Company itself was not averse to looting ships belonging to the Mogul emperors. These ships conveyed explorers, adventurers, merchants and occasionally plant collectors. Some of them eventually returned, others elected not to, some were never heard of again. The seventeenth, eighteenth and nineteenth centuries were the golden era of the descriptive naturalists, following close on the heels of the explorers, if they were not the explorers themselves. The perspicacity of some of these natural scientists was extraordinary. One was Rumphius, who became blind on the island of Amboyna in present-day Indonesia in 1670, when he was forty-three years old. This did not prevent him finishing his *Herbarium amboinens*, which took him another twenty years, and was published in the Netherlands in seven volumes in 1741–51, long after his death in 1706. Such vast compilations had no immediate practical application, but they made possible the work of the systematizers such as John Ray and his successor, the Swedish botanist Linnaeus, who founded the modern system for classification of plants.

The botanists were obsessed with cataloguing the products of creation, but they were also alive to the possible uses of the plants they found. Many of these seemed to have no particular use, some were poisonous, some seemed to be medicines. Among the bountiful products of nature reaching Europe from all corners of the globe were some with quite extraordinary properties.

Opium had been known for many years, knowledge of it reaching the West through the writings of the Arab physicians. It formed the basis of the apothecary's most powerful potions. When Charles II lay dying, he paid his personal physician Dr Jonathan Goddard £5,000 for his recipe for 'English drops'. The King found out that they consisted of opium mixed with distillate of human brains and skulls. He probably did not feel cheated, for, without the kind of proper scientific evaluation that was still a couple of centuries away, there was no way for him to know that only the opium content was of any use. Human heads were actually an established nostrum. The skulls of Irish convicts were especially prized. The moss (usnée) that grew on them as they rotted on the gibbet found its way into the pharmaceutical cabinets and thence into the first pharmacopoeias when they began to be published.

The number of drugs of real usefulness was strictly limited, but it was quite easy for the physicians and apothecaries to prescribe other drugs of dubious worth by claiming their specialist knowledge. They needed only one undisputed product as powerful as opium to be able to claim that the latest import from abroad was just as valuable in a different way.

By the eighteenth century, though, it was true that opium was not the only powerful and genuinely useful drug known. There was also the quinine-containing bark of the cinchona trees from the New World, for example. This came back to Europe through Spain and became known as the 'Jesuit bark'. Quinine is a good medicine against malaria, and a faint echo of its medicinal use lives on in the familiar gin and tonic; the 'tonic water' gets its bitter taste from quinine. But the history of quinine shows graphically how lack of scientific knowledge led to important mistakes, sometimes with tragic consequences.

First, it was incorrectly assumed that cinchona bark was effective against all fevers, not just malaria. Here the main defect was a lack of medical knowledge; the fact that malaria is caused by a mosquito-transmitted parasite would not be established for a long time. In the meantime, until medical studies proved otherwise, there did not seem any reason not to use cinchona bark against, say, pneumonia.

Secondly, no one knew what caused the bark to have its effect. Pure effective compounds had not been isolated from it; the techniques had not been developed. Here it was the chemistry that was lacking. The only thing that the physicians had to go on was that the bark was intensely bitter. Therefore, they assumed, perhaps *all* intensely bitter substances could cure fever. This incorrect idea lasted well into the nineteenth century, and there was a lot of fruitless effort put into making medicines against fevers from almost anything that was bitter – for example, coffee grounds.

In 1804 people lost their lives as a result of this ignorance.[4] For the previous twenty years or so, another bitter bark known as angostura had been imported from Venezuela and recommended as a medicine. It is now known that it does not contain any quinine and has no medical use, but it is not harmful and is still sold today for flavouring food and especially drinks. But in 1804 it was sold in apothecaries' shops to use against fevers. Remarkably, neither the apothecaries nor anyone else in Europe seemed to know for certain which plant the bark actually came from. There were various opinions.

In 1804 the apothecaries in London received, through the East India Company, a consignment of intensely bitter bark that superficially resembled angostura, but had a much more powerful, disgustingly bitter taste. The consignment was sold on to the Netherlands, where the apothecaries began adulterating batches of angostura with it and sending them all over Europe. Almost immediately, fatalities began to occur in Vienna, Riga, St Petersburg and other places, for the 'false angostura' bark was the bark of the poison nut tree *Strychnos nux-vomica* and contained large amounts of the deadly poison strychnine. Once again, it took years to establish what false angostura actually was. Some writers were confused, because true angostura came from the Americas while the *Strychnos* bark came from the Old World, and therefore scientific confusion between them should have been impossible. It did not occur to them that the apothecaries could have been dishonest enough just to mix them up. The correct identification was made just in time to prevent a large battalion of the British army in India being poisoned by it. So a combination of stupidity, greed and ignorance killed a four-year-old boy in Hamburg and many others.

By this era, there was some hope that the plant specimens being collected and sent back by the naturalists could not only be immaculately catalogued, but might be subjected to chemical investigation with some hope of useful results. But the standard of chemical investigation was still notable for its primitive nature.

A major expedition at this time was one that the famous explorer Alexander von Humboldt undertook with his friend Aimé Bonpland, who had given up a medical career in favour of his obsession with plants. Humboldt, a flamboyant homosexual, remembered meeting Bonpland in Paris: 'You know how it is, when you go out and hand in your key, you stop to have a few friendly words with the porter's wife. There I would often meet a young man carrying a botanist's vasculum. This was Bonpland.'[5] The two travelled in the American tropics for five years from 1799 to 1804.

> What a harvest of precious plants were offered to us by, on one side, the mountain ranges of New Andalusia, the valleys of Cumanacoa, Cocollar and the surroundings of the convent of Caripé, and on the other the immense plains and cultivated lands besides the forests of Guayana! . . . in these regions watered by incessant rains, the soil is covered by a multitude of unknown plants. The work of several centuries would not allow us to describe them all.[6]

The journey was not a picnic, though. The minute size of the canoes in which they spent whole months together, the burning climate, the multitude of venomous insects, the continuous rains, and (rather bathetically) the lack of paper can be appreciated only by those who have experienced them, Humboldt recorded. This did not prevent them collecting 60,000 specimens, which were sent back to Europe in three separate consignments for security reasons, one of which was lost when the ship sank.

As well as pressed botanical specimens from the many new species they encountered, they collected material for chemical examination. They discovered a new species of cinchona. The bark, held in the mouth for a while, gave rise to a strong astringent and aromatic flavour. Bonpland said that he thought the new species might be employed in medicines with success, and regretted that he had not collected enough material for it to be tested. But the uncertainties attached to deciding whether even one plant might really be useful could take decades of work by physicians to resolve. As Bonpland noted, even within this one species the appearance of the bark, and its bitterness, varied between the branches and the trunk, and whether the tree was an old one or a young one. And here they were finding hundreds of new species!

On Mount Quindiu in the Andes, they discovered a new kind of palm tree, which they named *Ceroxylon* because its trunk contained wax, or *cire*, which the locals used to make candles. They brought back a small flask containing a sample of the liquid found in the trunk cavities for the chemist Nicholas Vauquelin to analyse.[7]

Vauquelin, together with his slightly older associate Fourcroy,[8] had been one of France's leading chemists in the immediate post-Lavoisier era. He was of relatively humble origins and spoke no foreign languages, and it seems that the more aristocratic Fourcroy may have exploited Vauquelin's superior experimental skills to produce joint publications to which he had contributed little more than the background and phraseology. There were also rumours that Fourcroy had conspired to have Lavoisier out of the way, although these since seem to have been discounted. Napoleon made Fourcroy a count, but on the same day, 16 December 1809, he announced, 'Je suis mort', and fell down dead.[9]

As the discoverer of the fascinating new element chromium in 1780, Vauquelin had acquired great prestige and popularity with the revolutionary government. He had isolated urea from urine in 1799. His personal life in Paris was apparently extraordinary. Davy, visiting in 1813 found him, aged fifty, living in one room, which served as bedchamber and drawing room, with the deceased Fourcroy's two sisters, one of them sitting up in bed peeling truffles. Davy found his conversation shocking in its lewdness and tactlessness.[10]

Vauqulin's 'analysis' of the material from the new palm *Ceroxylon* was pathetic by comparison with the magnificence of the description and illustration of the plant itself by Bonpland. 'M. Vauqelin, to whom we sent a little flask of this product, was pleased to analyse it. He found that it consisted of two thirds resin and one third of a substance that was precipitated by alcohol and had all the properties of a wax. It was more brittle than beeswax. The results of this great chemist will be published in the annals of the Museum.'

Bonpland went on: 'Nature produces in a family of plants, and in the organs which appear to be very uniform, highly variable mixtures in which she is pleased to vary the combination of the elements and the mysterious interplay [*jeux mysterieux*] of their affinities.'[11] This was not saying very much. Bonpland was not a chemist, and could hardly be expected to realize how the analysis by his 'great chemist' would soon appear primitive.

The way forward to a more meticulous natural-product chemistry had already been shown by a better man than Vauquelin, some time before.

If there is one chemist (or rather, apothecary) who deserves to share a large part of the credit with Lavoisier for founding the modern subject, it is Carl Wilhelm Scheele, one of eleven children of a merchant from Stralsund on the south shores of the Baltic.[12] Not a great deal is known about him or his family; even his nationality is uncertain. At the time of his birth in December 1742, just a few months before Lavoisier, the region of Pomerania was part of Sweden, but since 1815 it has been German, and it is not clear which nationality Scheele's family claimed for themselves. Scheele himself wrote in both languages, but mostly Swedish.

He had such a retentive memory that it was said that, once he had read a book, he remembered all of it and never had to consult it again. After he had become employed as an apothecary in Gothenburg, the range of his researches was astonishing, until his death at the age of only forty-three. Davy later said of him: 'Nothing could repress the ardour of his mind, or damp the fire of his spirit . . . the example of Scheele demonstrates what great effects may be produced by small means when genius is assisted by industry.[12] Only one thing held him back; he was a convinced phlogistonist, and, because of his early death, he never really had the chance to follow his contemporaries in switching to Lavoisier's chemistry. Had he lived long enough to do so, his achievements might have been even greater.

Scheele is best known for finding oxygen before Priestley, but he did some other very remarkable things – for example, discovering pasteurization long before Pasteur did. He found that, if bottles of vinegar were heated in a bath of boiling water for half an hour and then sealed, they could be kept indefinitely, even if they contained air. He also discovered prussic acid, or hydrogen cyanide, which he obtained by distilling the pigment Prussian blue with dilute sulphuric acid. In an experiment unlikely ever to be repeated, he reported that it had 'a peculiar, not unpleasant smell; a taste which borders slightly on sweet and is somewhat heating in the mouth; also it provokes coughing.'[13]

It was his researches on the plant acids that paved the way to natural-product chemistry. Acids, to the eighteenth-century chemists, were substances with a sour taste that made plant dyes such as litmus change colour, and reacted with alkalis, or bases, such as lime and soda. Inorganic acids such as sulphuric, nitric and hydrochloric were well known. There were one or two acids vaguely known that are what we would now call organic (carbon containing), such as formic acid, the acid responsible for ant and nettle stings, which had been obtained in the seventeenth century by distilling ants.

What stands out when reading Scheele's accounts of his experiments is the way that, without any fuss, he began to use modern chemical methods of isolation, and the clarity with which he distinguished one compound from another. He isolated tartaric acid from wine lees, citric acid from lemon juice, oxalic acid from sorrel and rhubarb, uric acid from urine, gallic acid from oak galls and malic acid from apples. He showed that these were all different, and all his results have stood the test of time. None of them was a mistake, or a mixture, or the same as one of the others. Scheele also carried out some chemical transformations. He treated his 'acid of apples', or malic acid, with nitric acid and converted it into oxalic acid, and obtained the same product by nitric acid treatment of cane sugar, which is not an acid at all.

Another acid that he obtained was lactic acid from sour milk, which we shall meet again later. It is well known, he recorded, that milk becomes sour in summer, although he was unaware that this is due to the action of living bacteria on it. He filtered off the whey, and attempted to 'separate off all the heterogeneous things from the acid in order to obtain it as pure as possible'. We can see him in the act of making the transition from primitive to modern chemical manipulation. If this purification could take place by distillation, he says, it would be the shortest way, but it does not succeed because, although a small amount of acid passes over, almost all of it remains behind in the retort, and, if the heat is further increased, it is decomposed. Therefore he devised a remarkably sophisticated procedure involving preparing the calcium salt, decomposing this with oxalic acid, or 'acid of sugar', to produce the very insoluble calcium oxalate, filtering this off, evaporating the remaining liquid to a syrup, treating this with alcohol and then water, and distilling off the alcohol. 'The acid of milk remained behind in the retort as pure as, in my opinion, it can be obtained by the art of chemistry'.[14] In the words of Partington, 'Every chemist who has attempted researches will look over the record of Scheele's discoveries with astonishment and admiration.'[15]

Because of his early death in Sweden, far from Paris, Scheele's work for the time being remained something of a curiosity, not sufficiently appreciated for its true value. The French chemists could not believe that an apothecary in such a remote part of Europe could produce such reliable work, and for many years accused his lactic acid, for example, of being the same as acetic acid, the acid from wine vinegar, which it is not. Scheele had pointed to the way in which, if the apothecaries were to take a more scientific approach to their work and apply the careful methods of Lavoisier to what they had in their cabinets, remarkable things might be found. But the vast majority of them were still traders and their science remained essentially descriptive. Even Scheele, wearing his apothecary's hat, was not slow to point out that his citric acid, ground up with sugar, made a most satisfactory lemonade powder,[16] and many of the things in his apothecary's cabinet were still ill-defined quasi-chemicals, like 'Ley of blood'.

Some time after Scheele's death, however, some apothecaries became interested in trying to obtain purer and more powerful extracts of opium. Derosne and Séguin in Paris and Serteuner in Germany were able to isolate from it a crystalline substance that they called 'morphine'. In 1805 Serteuner found that the crystals they had obtained from the crude extract of opium were actually salts of morphine with acids.[17] Morphine appeared to be a 'vegetable alkali', what we would now call a base – a substance that reacts with acids to form salts and was thus analogous to the well-known inorganic alkalis such as soda and lime.

Serteuner's results were initially overlooked in Paris, but, when he republished them in a different journal several years later, the group around Vauquelin (it would be incorrect to call them his 'research group', since this concept did not yet exist) began to look for other examples. It was either Vauquelin or another colleague Gay-Lussac who first recognized that there seemed to be a whole new class of plant products characterized by their basicity, or ability to dissolve in, and form salts with, acids. The new substances, later christened 'alkaloids', within a few years became the hottest and most exciting chemical research topic in Paris, although by 1820, when the students there included the young Robert Christison, founder of forensic science in Britain, most of them considered Vauquelin himself something of a back number.[18]

Scheele and Lavoisier had indicated the route away from the furnace and towards the more careful methods of manipulation that were necessary to isolate and handle the delicate organic natural products. The vast majority of what was then known about chemistry concerned inorganic chemistry. It is hardly surprising, therefore, that the French scientists were biased towards treating their extracts with inorganic substances such as salts of lead and copper. Fourcroy carried out some early analyses of cinchona bark, but, although his methods were a shade less drastic than those of the alchemists (involving extraction with boiling water, treatment with inorganic salts and so on), they were still not refined enough and he isolated nothing of interest.[19]

One step forward came with the gradual realization of the importance of using suitable solvents. Foremost among these was ether, a light, mobile liquid with a pleasant smell, first discovered by the alchemist Conrad Gessner as long ago as 1552. Ether can be easily prepared by heating alcohol with a little sulphuric acid. It dissolves most organic substances readily. It is nearly immiscible with, and much lighter than, water. An ether extract containing the natural products can be washed with water to remove inorganic salts, the ether layer floating on top of the water from which it can be separated. It has a low boiling point, below blood heat, and can casily be evaporated, leaving behind the natural products without any need for strong heating. It has a number of drawbacks: the vapours are narcotic, and it is extremely inflammable. Many a chemical laboratory has been destroyed over the years by the deadly tendency of ether vapours to creep across the surface of a table or floor until they encounter a flame, then ignite. But ether became an invaluable ally to the French investigators. It was a substance well known to Vauquelin, and, during some investigations that he made into low temperatures, he was the first to solidify it. It eventually emerged that the alkaloids, these new acid-soluble plant products, could be purified by washing the ether extract, not with water, but with dilute acid. The alkaloids would now pass into the acid layer, leaving the other plant products, the oils, waxes and so forth, behind. After separating the two layers, addition of a weak alkali would neutralize the acid, and the alkaloid would separate out as a precipitate, or could be extracted into a fresh layer of ether. These techniques, simple but vital, were not discovered all at once.

At this period, Paris was a veritable rabbit warren of laboratories. It is often impossible to work out who was working where, and a great deal of time was spent walking between different laboratories and lecture rooms, often through insalubrious and dangerous districts. It would be a mistake to assume that all, or even most, of these laboratories were lavishly equipped.

Another associate of Lavoisier had been the apothecary Bertrand Pelletier, who, although dying at the age of thirty-six had filled several posts in the revolutionary government. He had sired a son, Pierre-Joseph Pelletier, born in Paris in 1788, who in his turn became a chemist and an associate of Vauquelin. Pierre-Joseph Pelletier would eventually become one of France's greatest scientists and would go on to isolate not only quinine from the South American Jesuit bark, incidentally founding a considerable business, but also chlorophyll from green leaves.

The second isolation of an alkaloid came in 1818 when Pelletier the younger and Joseph-Bienaimé Caventou isolated the deadly poison strychnine in pure form from St Ignatius beans, a poisonous seed from a plant closely related to the poison-nut tree.[20] First they extracted the beans with ether, then evaporated the solvent to obtain something resembling a thick greenish oil or butter. This substance, they admit, they had previously regarded as being the pure essence of the poisonous plant, but ideas about purity were beginning to change. Accordingly they treated the oil with several portions of boiling alcohol, filtering the extracts when hot to remove some insoluble matter and then again after cooling to remove a small amount of waxy material that had separated. They evaporated the alcohol and obtained a yellow-brown residue. They tried treating this with ether, with salts, and with oxides of various metals but were unable to effect a further purification. Finally they tried alkali, and obtained an abundant white crystalline precipitate, extremely bitter to the taste, that they recognized as the true active principle. In these beautiful but sinister crystals resided all the poisonous properties of the plant itself. This they recognized as the second member of the new class of alkaloids. In their paper describing their isolation they stated clearly, possibly for the first time, the scientific principles of plant drugs. The medicinal effects of a plant, they say, are due not to the plant as a whole but to one or more active ingredients.

Scheele had also shown the way to the most effective simple method for purifying organic compounds. He studied the important product 'flowers of benzoin' (benzoic acid) obtained from Gum Benjamin, and clearly described how to obtain a pure specimen of it. It is difficult to avoid the impression that Scheele was a more talented experimenter than most of the Parisian chemists of twenty years later. 'If it is desired to give [it] a shining appearance, it should be placed in enough water . . . and dissolved by gently boiling – after which it ought to be quickly filtered, hot, by means of a cloth, into a previously heated flask, when one has the gratification of seeing beautiful crystals shooting out as soon as the solution becomes cold.'[21]

Producing natural products in crystalline form would be for many years highly important. The beautiful outer symmetry of crystals reflects the purity of the substance within, because the molecules in the crystal are lined up in a three-dimensional lattice measuring millions of molecules on each side, each one (ideally) in exactly the same orientation. During the classical period of organic chemistry the aim of any investigator was always to try to get his material in a pure enough state for it to crystallize. Once this had been achieved, it could be further purified by the simple process of recrystallization. One takes the sample of the reasonably pure substance and dissolves it in a hot solvent, such as alcohol, water or chloroform. If the right solvent has been chosen, crystals appear, often spontaneously, sometimes with a little persuasion, sometimes with difficulty. Time is the most valuable ingredient of the process. Overnight perhaps, prisms or needles or rhombs appear on the glass walls of the flask or floating on the surface of the liquid. The recrystallization may have to be repeated several times until the compound is fully pure, but the beauty of the method is that impurities disappear from the crystals even if the impurity is less soluble in the solvent than the desired compound.

The operation of crystallization, so dear to the heart of the organic chemist, is seductive, and is partly responsible for the overemphasis on the tractable and the beautiful that characterized much of organic chemistry's classical period. Some chemists spent too much time on the elegant rather than the truly significant; nature's important molecules do not always crystallize very well. If a substance is too impure, it will not crystallize at all, and until the mid-twentieth century the methods available for purifying very impure products were strictly limited. Many of the important compounds with very small molecules, such as methane CH_4, are gases, and therefore much more difficult to study than crystalline compounds.

Once it was recognized that pure substances often crystallized well, there were things that could be measured to give an indication of genuine chemical purity, and this concept became much better developed. The crystal form could be carefully studied under a magnifying glass, and, later in the century, Louis Pasteur would make a remarkable discovery by doing just that.

Crystals of a pure substance generally melt instantaneously as the temperature is gradually raised. In the nineteenth century a small sample of the crystals, just enough to see clearly, was placed in a tiny glass tube held to the bulb of a thermometer with a piece of twine. The thermometer was lowered into a bath of oil and the oil gently heated. As the temperature of the crystals rises, the molecules in the lattice vibrate faster and faster until at a certain point the forces holding the crystal together are insufficient to buffer it against the vibration. This happens simultaneously throughout the crystal, which suddenly melts. The melting point was the best criterion of purity that a nineteenth-century chemist had. Impurities lower the melting point and make it indistinct. Even better, two samples can be shown to be identical by the method of mixed melting point. If two samples of crystals of the same substance are ground up finely together, the melting point will be unchanged. If, however, they are different, each will act as an impurity in the other and the melting point is depressed and broadened. (This, of course, shows that two samples are *identical*. It does not show what they *are*.) None of these methods is perfect. Strychnine, for example, does not have a sharp melting point; if heated slowly, it begins to melt at about 268°C, but, if heated instantaneously to just below 290°C, and then more slowly, it melts at this higher temperature. This behaviour indicates that at above 268°C or so it is starting to decompose with the heat, producing impurities. Strychnine is also polymorphic; that is, it can crystallize in six or seven different crystal forms, depending on the temperature and solvent. Polymorphism is quite a common phenomenon, so that neither melting points nor crystal form were completely reliable indicators of purity for the early chemists. Pelletier and Caventou also made a mistake when they said that strychnine contains only carbon, hydrogen and oxygen 'like the majority of vegetable substances'. Analysis was still primitive, and they had missed the fact that it contains nitrogen also, and it is in fact the presence of this element that leads to the basic properties of the alkaloids.

Other alkaloid isolations soon followed. French physicians, such as Fouquier at the Hôpital de la Charité, began to take the alkaloids seriously. Their use, together with other plant drugs, was popularized by François Magendie, usually regarded as the founder of the science of physiology, in his *Formulaire pour la préparation et l'emploi de plusieurs nouveaux medicaments* (1821), which soon went to five editions. Following on from the success of quinine, the powerful new plant drugs began to appear highly progressive compared with the older remedies of blistering and bleeding, although strychnine for one developed a wholly undeserved reputation as a useful drug, which it did not deserve. It killed a great many people before its eventual disappearance from the pharmacopoeias a century and a half later. Pelletier founded a laboratory in Paris for the commercial preparation of alkaloids, and his example was soon followed in London by Thomas Morson.

The cornucopia of Nature was beginning to open up to the scientists and physicians. But their knowledge was still empirical. Alkaloids contain nitrogen, morphine can be found in this poppy and not in that one, quinine has such and such a melting point, and so on. There was no explanation of these facts, no real theoretical background to chemistry. The first attempt to provide one awaited the arrival on the scene of a mild-mannered nonconformist teacher from the north of England.

FOUR

Alike, Globular and All of the Same Magnitude

One science only will one genius fit;
So vast is art, so narrow human wit . . .
        Alexander Pope, *An Essay on Criticism* (1711)

The alchemists and early chemists were not very good at making measurements. They quite understandably preferred explosions, poisons, things changing colour and the outside chance of making their fame and fortune. Taking careful measurements was not only boring, but they had no evidence that it would be of any use.

The idea that it might be worthwhile to measure things began with physics. Curiosity needed answers to eternal questions such as just how big is the sun and how far away? Measurement eventually spread to chemistry, but not all at once.

It is obvious to us now that the weight (or, more correctly, mass) of substances taking part in chemical reactions is of crucial importance. This was by no means clear to the early chemists. They believed that you could create matter by incantations, and they could point to numerous examples of things happening that seemed to support this view. When a tree grew, it apparently created itself from nothing, or at the very least from water, as Van Helmont thought he had proved. When you chopped it down and burnt it, it nearly all disappeared again. So the law of the conservation of mass, or indestructibility of matter, took a long time to emerge. It was Lavoisier who explicitly stated it, and finally disposed of any idea that matter could appear and disappear. Weighing then became the basis for working out where the matter goes when a chemical reaction takes place.

Some people knew this already, in an informal kind of way. For centuries past, rich people had been demanding information from assayers. How much gold does this sample of metal contain? Am I being swindled? It would not have done the court assayer a lot of good to venture the explanation that some of the gold had just disappeared overnight because an ill-intentioned enemy had pronounced a spell on it. By 1500 at the latest, this would not have worked. The assayers recognized the importance of weighing and developed ways of using crude balances to get tolerably accurate results. Gradually their knowledge spread from alloys to chemical compounds of the metals, like silver chloride. They developed look-up tables that could give them the answers without any knowledge of the underlying events. So, without really knowing it, they developed the foundations of chemistry.[1]

They realized that, if a substance was composed of two components, then this would always be true, and its properties would always be the same and could not be duplicated by a combination of some other set of components. From this it is (to us) only a short step to weighing the constituents carefully, determining their relative proportions and stating that the relative weights of the two components are always the same. But the emergence of these ideas took many years. The path was strewn with complications hardly noticed by the modern chemist.

There was still the problem of defining purity. This went well beyond the need to have pure substances, in the modern sense. There were more fundamental difficulties in defining what was meant by the term. The traders, metallurgists and apothecaries were well used to variations in the properties of their substances. If different samples of fuller's earth or indigo or opium had different attributes, then why not different samples of salt or zinc chloride? Variations in samples were considered the norm. Chemist A would describe his zinc chloride as if it were a natural curiosity, not expecting his sample to be exactly identical with that of chemist B. The fact that zinc chloride was always the same could be established only by careful measurements. But what, exactly, to measure? As far as organic chemistry was concerned, an important step came when the concept of a pure compound emerged (Pelletier's crystalline strychnine, not his greenish oil or butter). For some time before this, inorganic chemists had been trying to discover the rules that governed the combination of their simpler substances.

Shortly before 1630, Jean Rey, a physician from south-west France, or rather his friend Monsieur Brun, heated up 2 lb 6 oz of fine English tin in an open vessel for six hours, until it had all been converted into white calx. This calx (tin oxide) they found weighed 2 lb 13 oz. In the booklet that he published describing their experiment, Rey came very close to leaping over the next century and a half of confused speculation and carving his name on the altar of science in giant rather than rather small letters. For he stated quite clearly that the increase in weight comes from the air.

But he went further than that. No matter how long the calx was now heated, it got no heavier. His friend's 2 lb 6 oz of tin can absorb 7 oz in weight from the air and no more; if the heating is stopped before it has absorbed the 7 oz, it will be found that the substance in the container is now a mixture of tin and calx.

Why, he asked, should not the calx increase infinitely? Clearly, 'Nature in her inscrutable wisdom has set limits to which she never oversteps.'[2] Unfortunately for Rey, the explanation that he wheeled out was not the right one. It arises, he said, from the fact that tin and calx are solids, and air a fluid. In the same way that a barrel of sand can absorb so much water and no more, the calx becomes saturated with air. It was not like mixing the good wine of Gascony with water, where one can obtain any mixture that one wants because they are both liquids.

Rey had stumbled upon, although not correctly explained, what later became known as the law of constant composition, which was not clearly stated for approximately another 170 years in the era of Lavoisier's new chemistry. *A particular chemical compound always contains the same elements united together in the same proportions.* Cinnabar from Japan has the same ratio of mercury and sulphur as cinnabar from Almaden, said Proust, whose law it is. In all the known parts of the world, you will not find two muriates of soda (sodium chloride or common salt). According to my view, a compound is a privileged product to which nature has assigned a fixed composition . . . between pole and pole compounds are identical in composition . . . in all the world there is but one saltpetre; one calcium sulphate; and one barium sulphate. Analysis confirms these facts at every step.[3]

Another significant empirical law was the law of multiple proportions. Sometimes, two elements may combine to form more than one compound; for example, there are two copper oxides and two iron chlorides. But, if they do, the combining powers increase by leaps, not gradually, and the ratio of the weights of the elements in the two compounds is always a simple ratio; for example, the amount of oxygen in the black oxide of copper is twice that in the pink oxide. The law of multiple proportions requires that the law of constant composition be true.

These laws result from the fact that matter is composed of atoms. When Rey and his friend M. Brun heated their tin, each atom of tin combines with two oxygen atoms, no more and no less, to form tin oxide, SnO_2. If insufficient air was used to oxidize all the tin, what was left in the pot was a mixture of SnO_2 and unchanged tin; it was not something intermediate between tin and tin oxide.

We have to make two major qualifications about the implications of Rey's experiment. Both of them were to haunt the new science of chemistry, when it emerged, for many years. The first is this: although the existence of atoms is the explanation for the law of constant combination, it is not the *only possible* explanation. Atoms explain Proust's law, but Proust's law does not prove that atoms exist.

The second qualification is equally vital, and caused even more problems than the first one. Even if Rey had known for certain about atoms, neither he nor anyone else would have had any way of knowing how many atoms of oxygen were combining with each atom of his tin. We shall return to this.

One of the factors though that seemed to distinguish the new natural products from the ordinary inorganic chemistry of water, salt and tin oxide was that the law of multiple proportions did not usually operate in a simple way. When sufficiently accurate analytical methods were developed to measure the weights of the elements present in them, it was usually found that these were not simple small numbers. For example, the mass of hydrogen combining with a given mass of carbon in quinine, expressed as a ratio of that combining with the same mass of carbon in morphine, is $102:95$. This is hardly a simple number and, given the inaccuracies of early analytical techniques, one that was almost impossible to measure sufficiently accurately. Until some talented analysts put in an appearance, this seemed another reason to regard organic compounds as special.

But, for the run of everyday inorganic substances, the simple ratios held sway. How could they be explained, and did they point to some simple underlying arithmetical regularities in the organization of nature?

John Dalton was born in 1766 near Cockermouth in Cumbria.[4] He spent virtually the whole of his life quietly in Manchester, something of a recluse pursuing a lifestyle that was by any standards dull, punctuated by half-pints of beer, pipes of tobacco and a game of bowls every Thursday at the Dog and Partridge Inn. Later on he became widely honoured and was much in demand as a lecturer in London, Paris and Glasgow, despite his highly homespun manner. He continued to speak in the Cumberland dialect and at one point caused offence in London by referring to the elements as 'articles'. French worthies from the Académie des Sciences, visiting him after the publication of his atomic theory, were said to have been astonished by his modesty and lack of pretension.

He had little formal education, but was interested in many aspects of science or 'natural philosophy'. He studied meteorology and thus the properties of air and other gases, and of water and other liquids. Colour-blindness, still called *daltonisme* in France, was one of his great discoveries; one of Dalton's eyes is preserved in a jar at Manchester University.

The democratic spirit engendered by the rise of nonconformism was a major contributory factor in allowing Dalton to flourish, and the north of England was to become a hotbed of questioning scientists, of whom Dalton was among the first. The local Quaker community developed his talents. The circumstances of society in late eighteenth-century England had allowed this intelligent but half-educated layman to become the author of major scientific papers, to get them published and for them to be accepted by his scientific contemporaries.

By Dalton's time, the concept of an element, fully developed by Lavoisier, had been around for twenty or thirty years. But the concept of atoms does not automatically follow from that of elements. To us, it seems such a short step to say that, if the weight of two elements in a compound is always constant, this is because the elements are composed of atoms. But this is with the benefit of nearly two centuries of experience. In Dalton's time the well-established fact that there are chemical elements did not require them to be made of atoms.

It has been said that Dalton did not discover the atomic theory, he invented it. It would be grossly incorrect to say that he wrote down the rules of chemical combination, especially the law of constant composition, and then devised his atomic theory to explain them. In 1810 the principles were not fully established; as the century progressed, the atomic theory came to be seen, little by little, and by the majority of chemists only, as the best working hypothesis by which these phenomena could be explained. But it had its ups and downs, to put it mildly.

The idea that matter is not infinitely divisible goes back to the Greeks, although their level of discussion was philosophical, not experimentally based. Democritus (fifth century BC) introduced atoms as a concept in order to reconcile the dilemma that, if the universe was constantly changing, which it demonstrably is, how could there be any sure knowledge, if there was nothing unchanging to be known? According to Democritus, this was because the atoms themselves were unchanging, and all changes in the universe were due to changes in their arrangement. This is not far from the true state of affairs. But these ideas raised as many difficulties as they eliminated. For example, presumably the atoms had spaces between them allowing them to be rearranged; but, if this space had nothing in it, how could it exist? The Greeks were puzzled. One wonders what they would have thought had they been confronted with one of Torricelli's tubes containing a vacuum. Would they have been satisfied that his 'visible' empty space was the same thing as the invisibly small nothing between their invisibly small atoms?

It may seem facile to write off the next two millennia of thought on the subject of atoms, but in truth this is more or less where the matter rested until the time of Dalton. Indian philosophers debated atoms at length during what we call the Middle Ages, but the question remained a puzzle. There were arguments for and against. For example, if there was no such thing as an atom, they reasoned, then a mustard seed would have an infinite number of parts, and should fill up the whole earth.[5]

This is what Dalton postulated about his atoms. The elements, he said, are composed of tiny particles called atoms. They are indivisible and indestructible. All the atoms of a single element are the same, and have the same mass, but the atoms of different elements are different. Chemical compounds consist of the atoms of different elements joined together. Compounds have their own fixed composition because they contain a fixed ratio of atoms, and chemical reactions involve the rearrangement of those atoms.

Dalton never seemed clear about how the idea of atoms came to him.[6] The first explanation was published by Thomas Thomson in a book of 1830. He wrote that Dalton had told him that the concept arose from some studies of hydrocarbon gases that he had carried out from 1802 onwards. These established that methane, or 'carburetted hydrogen' (modern formula CH_4), contained exactly twice as much hydrogen as ethylene, or 'olefiant gas' (modern formula C_2H_4). In studying these simple hydrocarbons methane and ethylene, Dalton was placing himself at the heart of the territory that needed to be thoroughly explored in order to explain chemical combination in organic substances. These are the molecules that appear in the first chapter of every high-school organic chemistry text, and it is frustrating to see how they were largely sidelined for the next half-century.

In the first biography of Dalton, written in 1854, the author said that Dalton had told him about receiving inspiration from some excellent tables published by Richter, who in 1791 had determined the weights of various acids reacting with various bases, giving their 'equivalent weights'. There seems some doubt as to whether Dalton in fact saw these tables before the publication of his hypothesis, or whether he rationalized his insights when he saw them subsequently.

Then, in 1896, both of these explanations were contradicted by the discovery of Dalton's own notebooks in the meeting rooms of the Manchester Literary and Philosophical Society, together with his other unpublished manuscripts. A.N. Meldrum studied them carefully and came to the conclusion that the crucial experiments had to do with discovering that nitric oxide would combine with oxygen in two ways, differing in the ratio 1:2, and nothing in between.

So there are now three explanations, but they are at least related in the sense that they arise from the results of chemical experiments, as we might hope for from a hypothesis intended to explain chemistry. But Dalton himself, in a Royal Institution lecture of 1810 (the notes for which also surfaced in 1896), gave yet another explanation – one based on physics not chemistry, and on incorrect physics to boot! His atomic theory came to him, he says, from a study of the diffusion of gases, which he could explain only on the assumption that the particles of different gases were of different size. So now there are four explanations! But the dates cited by Dalton are not consistent.[7]

Most of the 1896 material was destroyed by bombing in 1940, but people still debate these various explanations. Dalton shows how the mental processes leading to new theories are often irrational and instinctive. The process may be unclear even to the conscious mind of the instigator. There are some superficial resemblances to the uncertainties surrounding Kekulé's claimed inspiration later in the century.

In addition to not being clear in his own mind about how he got to his theory, Dalton never seems to have been clear whether he literally thought that all matter in the universe was composed of indestructible atoms. It could have been a theory only of 'chemical atoms' explaining chemical reactions and not much else. The Swedish chemist Berzelius, geographically isolated, took some time to hear rumours of the atomic theory. When he did learn of it, it struck him with considerable force, but when he received a copy of Dalton's *New System* in 1812 he was disappointed; even the mathematical parts contained errors, and Dalton wrote like someone with a 'preoccupied mind', or like an 'absent-minded person' (*tête préoccupée*).[8]

Although Dalton lived until 1844, in the second half of his life he did little of scientific interest. He continued to lecture on his atomic hypothesis, but did not really develop it, or be aware of the territory potentially opened up by it. It sprang into existence from his fertile brain, and that was more or less it. This has partly to do with Dalton's own leanings as a natural philosopher. The boundaries between physics and chemistry had not yet been defined, but by inclination Dalton was really a physicist, interested in his meteorology and atmospherics. His knowledge of laboratory chemistry was patchy and he was not by inclination an experimental chemist. In one way this was an advantage, because, unlike most chemists at the time, he was not troubled by doubts about Lavoisier's list of elements, but took it as a given.[9] It is idle to speculate what Dalton might have dreamed up had he spent a year or two working in an apothecary's shop. But such an outward-looking departure was not in his nature. The German chemist Mitscherlich, himself not the easiest of personalities, visited England in 1824 and found Dalton 'short and stooped and taking so little interest in everything one tries to discuss with him that the time I spent in his company was the most awkward of the entire trip'.[10] A new territory was opening with the exploration of the products of nature's bounty, but Dalton did not enter it.

The more one thinks about Dalton's action in getting his friend Peter Ewart to make wooden models of the atoms for him, the stranger it seems. What, precisely, did Dalton know or suspect about the properties of his atoms, and what could the models conceivably represent?

Imagine you are in an audience hearing an eminent astronomer talking about distant planets. He announces that careful measurements of the light of a star many hundreds of light years away have led to the conclusion that there must be a planet circling it. He then produces an orange-painted ball of wood three inches in diameter. 'And this', he announces, 'is a model of that planet'.

The members of the audience immediately begin to downgrade their assessment of his eminence. What is the point of this model? 'How big is the planet?', someone asks, and he has to admit that he has no idea. 'What is the significance of the orange colour?', someone else enquires. He replies that it has no significance, since he really has no idea what colour the planet might be. Surely he is not implying that the planet is made of wood? And yet modern astronomers actually know more about such a planet at the limit of their observation than Dalton knew about the atom. They know that it must be very big, otherwise it would be undetectable; it must be spherical and probably made up of ammonia, hydrogen and other gases, and so on.

Dalton had no idea of dimensions, either relative or absolute. Assuming that atoms did indeed exist (a necessary caveat until the end of the nineteenth century), there was to remain absolutely no clue about their size. They were too small to be seen through the best microscopes, and that set the upper limit; that was all. Writing in 1851, Liebig was to say: 'A piece of cinnabar large enough to be manifest to our senses must contain, perhaps, millions of atoms.'[11] If, for the sake of argument, the smallest amount the human eye can see is a microgram, we now know that the number of molecules of cinnabar (mercuric sulphide) in this barely visible speck is 2,700,000,000,000,000; his guess was a billion or more times too low.

Nor was there any information concerning the atoms' relative sizes or shapes. Dalton's wooden model atoms were made all the same size, but for all he knew the oxygen atom might be a million times bigger (or smaller) than that of hydrogen. In fact, the main piece of evidence he might have had to go on was that he thought the oxygen atom eight times heavier (more massive) than that of hydrogen; but he did not make it eight times bigger in his models. Some time *after* he first put forward the atomic theory, he began to think that some of the properties of gases might be attributable to differences in the sizes of their atoms. But in other places in his writings he seems to represent the atoms as points, having position but no dimensions. He vacillated, but eventually settled on saying that the atoms are all the same size.

> It is my opinion that the simple atoms are *alike*, *globular*, and all of the same *magnitude*, or bulk . . . my friend Mr Ewart, at my suggestion, made me a number of equal balls, about an inch in diameter, about 30 years ago; they have been in use ever since, I occasionally show them to my pupils. . . . I had no idea at that time that the atoms were all of a bulk, but for the sake of illustration I had them made alike.[12]

The Greeks had attributed many of the properties of substances to the differing shapes of their atoms. For all Dalton knew, the atom of oxygen was shaped like the leaning tower of Pisa, and that of zinc like the Taj Mahal; so why spheres?

Dalton's choice of model implies that, when atoms interact, they do not interpenetrate. Hard wooden spheres do not merge into one body when brought together, but lie side by side. For the atoms, this is not self-evident. Dalton could have dreamed up an alternative atomicity in which, when combination takes place, the hydrogen atom penetrates the oxygen atom, or vice versa, like a sperm entering an egg, to produce another body (sphere?) with different properties. He rejected that. This 'interpenetration' model was later toyed with by Berzelius, who said that it could arise from the dynamic system of German philosophy in which the universe is in balance between attractive and repulsive forces. It resurfaced at various times during the century. Interpenetration did not require the atomic theory. The elements could be nonatomic and infinitely divisible, and interpenetrate each other when they combined, as many of the alchemists thought.

There is one other consequence of Dalton's model. If atoms are discrete spheres, when they approach each other, all space becomes roughly divided into two regions: the region between them and the rest of space that is not between them. The way was potentially cleared for thinking about what goes on in the first region, but it was to remain a virtually unexplored place for many years.

Dalton's provision of his atomic models with hooks so that they could be joined together was also a remarkable thing. This question had already been addressed by the Indian philosophers.[13] If a thing is joined to another, it must have *parts*. How could it have parts if it was the smallest possible entity? Dalton seemed unaware of this philosophical dilemma, or, if he was not, he seems to have ignored it.

Dalton's atomic theory failed to take proper scientific root. It was premature, for the amount of firm evidence supporting it at the time that Dalton dreamed it up was negligible. The main justification for the theory was that it seemed to provide the most likely explanation of Proust's law and the other simple arithmetical laws governing the formation of chemical compounds. But it is a fallacy to state, as is often stated, that the atomic theory *proves* the laws of multiple proportion and constant composition. This is a logical error.[14] Some other explanation could have been possible. Over the following decades, chemists got into a fearful muddle trying to apply the theory widely without getting certain funda-mentals right. The contradictory results that surfaced strengthened the hand of the sceptics.

A paradox is that, before Dalton, most people assumed that matter was made up of atoms in some shape or form, but there had been no great interest in the question. After Dalton, everyone was more or less obliged to have an opinion. It was like the question of the existence of God; no one doubted it until people began to try to prove it in the eighteenth century, when the arguments against began to appear.[15] As with religion, you could be a believer, an unbeliever or an agnostic about atoms. The agnostic camp shaded off into unbelief for those, like Davy, who ignored the atomic theory because it had no predictive value. Davy's attitude was shared by many of his countrymen. England was the home of practical men and empirical philosophers; chemistry was an interesting and sometimes exciting new occupation, but who wanted to speculate about a theory that could not be proved either one way or the other? Davy thought that the law of constant composition was useful, but that all statements about atoms were hypothetical and useless. This was a widespread and justifiable attitude, certainly in the 1820s.

The camp of true disbelievers included positivist natural philosophers who felt that atoms were an unnecessary intrusion into the purpose of science, which was to explain the universe in mathematical terms. They spent much effort trying to evolve mathematical models for the behaviour of matter that did not involve atoms. These people had an Indian summer of fame in the 1860s, and we shall meet them again later.

For most of the nineteenth century a high degree of scepticism about atoms persisted. In France the influential Marcellin Berthelot, the author of 1,600 research papers and five times elected president of the French Chemical Society, did not believe in them, although it appears that, by 1860, a bare majority of the members of the society were opposed to his views.[16] When chairs in chemistry were created at the École de Pharmacie in 1859 and at the Collège de France in 1865, Berthelot held both of them. He later became Inspector General of Higher Education in 1876, Minister of Public Instruction in 1886 and secretary of the Academy of Sciences in 1889. Later still, he became France's foreign minister and was uniquely buried in the Pantheon next to his wife, who died on the same day. It is said that when he worked in the laboratory he wore a leather apron with a piece cut out to show the rosette of the *Légion d'Honneur*. Such a man could wield immense power over the dissemination of ideas. The commemorative volume produced in 1927 on the centenary of his birth must be the most pompous posthumous tribute ever paid to a scientist,[17] despite the fact that his chemical researches were primitive.

Even as late as 1904 the physical chemist Ostwald gave a lecture in which he famously tried to deduce all the laws of chemistry without using the atomic theory – an intellectual *tour de force* but by then a doomed one. A writer in the same year wrote: 'It is sometimes said that a theory of continuous matter is "Inconceivable". Those who say so may be supposed to know their own minds best. There is, nevertheless, the possibility that the theory of the future may not regard matter as atomic.'[18] This was the last gasp of even half-hearted scepticism, though. In 1909 Ostwald conceded that recent work in physics since the discovery of radioactive decay showed that atoms, and indeed subatomic particles, truly existed. But, altogether, the era of gradually decreasing uncertainty about atoms lasted for nearly a century.

The atom, according to Dalton, was the smallest fragment of an element that could exist. What about compounds: water, tartaric acid, quinine? There might be something analogous to the atom, the smallest particle of a compound that could exist without it being decomposed into something else. This would one day be called a molecule. *If* atoms existed, then they *might* combine together into molecules. *If* molecules existed, then they might have a life of their own. But these were very big ifs.

FIVE

The Flaming Intellectual Hearth

When a man is young he is so wild he is insufferable. When he is old, he plays the saint and is insufferable again.

Nikolai Gogol, *Gamblers* (1842)

When I was a child in 1950s London, my father and I would often visit the Portobello Road street market on Saturdays. One of the attractions there was an ancient seller of quack medicines, who would harangue the crowd in an attempt to get them to abandon modern drugs for his tried and tested nostrums.

His favourite demonstration involved treating the pavement as a kind of open-air laboratory. On the kerbstone he would place a substantial pile of potassium permanganate crystals. Making a well in the centre, he would throw in a handful of aspirins, then a lake of glycerine. After a minute or so the mixture would catch fire spontaneously – the powerful oxidizing agent potassium permanganate reacting vigorously with the glycerine. 'No, not yet,' he would say as the flames spread to the aspirin tablets. When it reached them, they would ignite and burn vigorously, leaving behind a sticky black carbonaceous residue. The peddler would then contemptuously dispatch the still-smouldering heap into the gutter with a sweep of his foot. 'There, you see?', he would proclaim. 'That's the sort of muck you are putting in your stomach.'

Quite what he thought he was proving by this firework display I have no idea. Aspirin tablets were largely mostly starch anyway. To me it seemed to prove two things, neither of which was what he had in mind; first, that nothing much had changed since Paracelsus, and, secondly, that chemistry is a fascinating subject.

Thus it must have seemed to the young Justus Liebig one early nineteenth-century day in the marketplace at Darmstadt when he saw a peddler make 'fulminating silver', or silver fulminate, a white powder that explodes with a loud bang when hit with a hammer. His fascination with this demonstration not only generated his early interest in chemistry; it formed a determination to find out more about the extraordinary fulminate. His experiments on this substance were to bring him into contact with his near-neighbour and near-contemporary Friedrich Wöhler. These two men, born within a few years and a few miles of each other in provincial, pre-Bismarckian Germany, came to personify the emergence of organic chemistry. Over a period of thirty years from about 1824, the two between them researched in virtually every area of the new science, published hundreds of papers and educated thousands of students (8,000 in Wöhler's case). Some of their early work together, such as that on fulminates and on benzene and its relatives (to be described in Chapter 9), included key experiments and led to vital new concepts.

Temperamentally they were completely different. Wöhler, the elder by just under three years and the son of a court official from Eschersheim near Frankfurt, was much the closer to the conventional idea of a scientist: precociously intelligent, self-educated, hard-working and well organized. He was quiet and modest with a dry sense of humour, slender and of very youthful appearance; when he visited England in 1835, he met Faraday, who thought he was talking to Wöhler's son. (Later, like all nineteenth-century professors, he grew a mass of facial hair and looked far less amiable.)

Wöhler was a constant counterbalance to Liebig, who became the more famous and influenced the development of science to the greater degree. Liebig has often been called 'the Father of Organic Chemistry'. If so, this irascible man, so fallible, was an immensely human parent, the antithesis of the great cold-blooded scientist Lavoisier. In a career punctuated by more than its fair share of failures, partial failures, major mistakes and phenomenal disputes, several of his achievements opened a new era in the unravelling of the secrets of the molecule.

Liebig was born in May 1803 in Darmstadt, at that time the principal town of the poverty-stricken ducal statelet of Hessen-Darmstadt.[1] His father was a hardware merchant and manufacturer of household products. During Liebig's childhood, the Grand Duchy, allied with France, suffered economically, first because of the British blockade and then in the aftermath by having been on the losing side. The peasantry was impoverished and the Liebigs were not affluent. Liebig's own ambition had at first been to expand his father's business into a successful chemical manufactory; when asked at school what career he had chosen, he famously replied, 'I want to be a chemist', and the rest of the class were convulsed with laughter. But his university experiences at Bonn and Erlangen, followed especially by his sojourn in Paris, made him think in wider terms and to realize the true potential of the new science.

The overriding characteristic of Liebig throughout his life was his immense and unremitting energy. Like all workaholics, he was constantly in despair about overwork but hardly ceased until his death at the age of almost seventy. The headstrong episodes of his youth were forerunners of the pig-headedness that was often to get him into trouble later when he was famous. This combativeness seemed in a way to exist outside himself, for the universal opinion of all who knew him personally was his geniality. 'He was not exactly what is called a fluent speaker, but there was an earnestness, an enthusiasm in all he said, which irresistibly carried away the listener,' one of his most famous students, Hofmann, said, remembering him years later. 'We still, in our manhood now, remember how ready we were as Liebig's young companions in arms, to make any attack at his bidding, and follow wherever he led.'[2]

In the first half of the century travel around Europe remained slow and costly. When Wöhler travelled from Germany to Stockholm in 1823 to begin work with Berzelius, he had to wait six weeks at Lübeck for the sailing ship on which he had booked passage. (He passed the time working in a laboratory making useful chemicals to present to the master on his arrival.) The highly efficient postal services were the principal means of communication. ('Just write "Chemical research" on the envelope and it will be carried free of charge throughout Germany,' Liebig wrote to a correspondent in 1839). It is in his letters and articles, of which thousands survive, that we see the argumentative, and frequently wrong, Liebig. His friend and collaborator Wöhler, a constant, though often ineffectual, counterweight, once wrote to him, 'Can one teach an ox to have understanding?' Liebig himself found these disputes stimulating; 'My mill has ever received its best supply of water from my opponents,' he once said. Much later in life, when he had moved from pure organic chemistry into biochemistry, agricultural chemistry and nutrition, the power of the ox was hardly diminished. In 1862 his English publisher refused to handle part of the new edition of Liebig's standard work *Agricultural Chemistry* because of the libels against various agriculturalists that it contained, and even destroyed his only copy.[3] He made real enemies; one, the Dutchman G.J. Mulder, was so upset by Liebig's insults about some work he had done on proteins that he published a 122-page book of invective against the famous chemist's treatment of all his opponents. 'Liebig respects *nobody's* name, moral character, social position, life or health,' he said.[4]

Even England, formerly respected, felt the lash of his pen by the 1850s. In a passage widely reported in British newspapers, he compared Great Britain to a 'vampire hanging on the breast of Europe . . . sucking its life-blood without any real necessity or permanent gain for itself'. This was before the invention of artificial fertilizers, and Liebig was referring specifically to the way in which Britain was using up the world's supply of natural manures, by methods including buying up the contents of Sicilian catacombs and grinding up the bones for application to agricultural land. In this nineteenth-century precursor to twenty-first-century ecological doom scenarios, Liebig predicted the inevitability of wars between the nation states over such resources.

Portraits of Liebig as a young man show a handsome dark-complexioned young blade (there was probably Jewish ancestry on his mother's side). At university, he became involved with the student drinking bodies that were the main focus throughout the country for subversive politics and revolution, and indeed being Liebig he had to go as far as to found his own corps.

The prevailing philosophical spirit in early nineteenth-century Germany was the *Naturphilosophie* of Goethe and Schiller. During the Biedermeier period of 1815–48, there was a reaction against the wilder aspirations of the *Naturphilosophie*, but the belief in the harmonic accord of nature remained. *Naturphilosophie* taught a romantic, self-oriented belief in experience and feeling, emphasizing individualism in a way that was indirectly a challenge to the authorities, who were constantly on the lookout for serious manifestations of the revolutionary spirit of 1789. German students were a byword for dishevelled free-thinking; hippies before their time. At the 1822 New Year celebrations there was drunken rioting lasting for several weeks, with Liebig as one of the ringleaders, among other things knocking off the hat of a local dignitary. In the paternalistic and authoritarian society of the small German states such riots were more significant than mere hooliganism. Liebig's immediate punishment was three days in prison, but his activities led to a search of his rooms, which found evidence that he was coordinating the demonstrations with students in other parts of Germany.[5] Such subversion could often lead to exile abroad or other severe penalties.

The Erlangen authorities summoned him to appear before them on 20 March 1822. What took place at the interview is not known, but reading between the lines it appears that Liebig may have taken fright, given a guarantee of good conduct and agreed to return to Darmstadt without a degree and refrain from political activity. At the time Kastner, the Professor of Chemistry at the university, was petitioning the ruler of Hessen-Darmstadt, Grand Duke Ludwig II, for a scholarship for Liebig to go to Paris. In such an environment where all depended on patronage, it is impossible not to imagine Kastner pointing out to Liebig how foolish he had been, and extracting from him certain undertakings for transmission to the authorities. Whether these undertakings included agreeing to rat on his fellow demonstrators is not known, although again there is circumstantial evidence. If so, this was the most disreputable episode in a life of reasonable probity. It has to be admitted that later, when he went into business, Liebig and his associates were not averse to a certain amount of sleight of hand in the mid-Victorian manner to increase the profitability of their ventures. In the mid-1840s he and Hofmann discovered a way of transforming the virtually worthless quinidine, available as a byproduct of quinine manufacture, into quinine. In order to corner the market in quinidine, they surreptitiously imported it into Liverpool and Hull described as 'resin' or 'asphalt'. One of the doctors involved seems to have tried out the new semisynthetic quinine by supplying it to London hospitals without telling anyone.[6]

Liebig's youthful indiscretions did not end with the rioting, though. At almost the same time, he had declared his love for the promiscuous poet August Graf von Platen. Later Victorian biographers were able to pass this off as the kind of intense same-sex relationship that was almost compulsory during the Romantic era and which may or may not have been sublimated later in the century into intense non-physical relationships of the 'comrades-in-arms' variety. Discovering the truth about nineteenth-century intimate relations is notoriously difficult, but it seems certain that Platen was deeply in love with the handsome Liebig and that Liebig encouraged him. Platen wrote Liebig six sonnets owing much to Shakespeare in style, and in 1825 Liebig felt able to write to him saying 'I embrace and kiss you and love you wholeheartedly'. But the very same letter announced to Platen his engagement to Henriette Moldenhauer. Liebig's heterosexual qualifications do not seem to be in doubt; earlier, he had told Platen that the reason he had left Erlangen in 1822 was that he had been conducting an affair with a married woman and her husband had found out.[7] The truth about this will doubtless never be established. Perhaps it was a device simultaneously to cool off Platen while drawing the scent away from suspicion of having betrayed his friends.

During the interval leading up to this 1825 letter, Liebig had been in Paris and had undergone a psychological sea change. He realized that his chosen career offered even more opportunities than he had formerly thought if he were to take it seriously enough and settle down. He saw classes of 400 students and more learning chemistry, and the French lecturers taught him the importance of a quantitative, or mathematical, approach to his work. The subject had a bright future that had not been so visible from the woods and decayed castles of Hesse. His early interests in pharmaceutical and chemical manufacturing, inherited from his father, grew into something more deeply scientific. A quarter-century after Lavoisier's execution, Paris remained the world's main centre of research, and a number of talented chemists worked there. We have already heard of Vauquelin, Fourcroy, Pelletier and others isolating the extraordinary alkaloids; Liebig worked in Vauquelin's laboratory for a time. Another famous Parisian chemist had a particularly strong influence on him.

This was Joseph Louis Gay-Lussac (1778–1850), who after working with Vauquelin and others became Professor of Chemistry first at the École Polytechnique, then at the Jardin des Plantes. Although, like Lavoisier, he was later described as rather cold and reserved, this was probably the sort of demeanour expected of someone who had become a fairly senior French government figure. He was a superlative lecturer and an enterprising experimenter. Christison described hearing him lecture to 1,200 students at the Jardin des Plantes.[8] One of his earlier researches consisted of going up alone in a hydrogen balloon to 7,000 metres above Paris on 15 September 1804, collecting air samples and showing that the chemical composition of the air was the same as at ground level. There was a friendly rivalry in these researches with von Humboldt, who seemed to crop up everywhere, and with whom Gay-Lussac shared a room on the Left Bank.[9] They seem to have had an agreeable relationship, complete with jokes. Once they needed to import some thin-walled glass tubes, on which there was a high rate of duty, from Germany. They wrote to the manufacturer asking him to seal the ends of the tubes and send them to Paris labelled 'Handle with care; German air'. The French customs men could not find any duty rate for air in their manual, and let the tubes in free.[10] Liebig described meeting von Humboldt in Paris during the summer of 1823, when he was twenty and Humboldt fifty-three. According to Liebig's version, Humbolt invited him to dinner and Liebig, not knowing who he was, failed to turn up; but perhaps he was uncertain whether he wanted to take up the invitation. Nevertheless, he eventually went, and formed a firm friendship.

Gay-Lussac, in collaboration with his co-professor L.J. Thenard, was the first to establish a fairly reliable method for analysing organic substances, in other words to find out what proportions of the various elements they contained. This was a problem of the greatest difficulty to the early chemists, requiring utmost experimental skill, but without it no theoretical advances could be made.

Early analyses of organic matter were crude and based on the methods of the alchemist handed down through successive generations of apothecaries. They relied on the standby of 'If in doubt, heat it up in a furnace'. The results obtained by this technique of 'destructive distillation' were usually useless, which did not prevent many hundreds of experiments being carried out. We have seen that by 1695 Wilhelm Homberg was already stating that such analyses applied to plant materials were of no worth, because every plant tried gave the same result.

The problem was that merely applying strong heat to an organic substance is not destructive *enough*. Complex series of decompositions take place resulting in various products boiling off at different temperatures in the form of mixtures. Such distillations resulted in so much oil, so much wax, so much water and so on; so what? Heating up single pure organic substances in this way could sometimes give interesting results, but they were usually impossible to explain. The term 'analysis' was a misnomer applied to such crude processes. Analysis means reducing to simpler components that can be understood, not the reverse. In Lavoisier's chemistry, true analysis meant finding out how much carbon, how much oxygen and so on were present in each substance. The elements present in the sample needed to be converted under controlled conditions, each to a single product that could be weighed or measured.

The method that Gay-Lussac and Thenard evolved was to destroy the sample with an excess of a strong oxidizing agent. They used the newly discovered potassium chlorate as oxidizer. A sample of the substance to be analysed was ground up with potassium chlorate and the mixture made into pellets. These were dropped into a carefully designed glass apparatus, strongly heated with a spirit burner, and the gases evolved collected. In this procedure the carbon content of the sample was all converted to carbon dioxide. The hydrogen content was assumed to have been converted to water, which was not directly collected but calculated by subtraction. Given the relative inaccuracy of measuring volumes of gases, their results were surprisingly accurate. They analysed five of Scheele's organic acids – acetic, citric, tartaric, mucic and oxalic – as well as sugar, woods, wax, olive oil, gelatine and other organic products, many of them not single compounds. In the case of sugars (milk sugar and cane sugar) and starch, they found that the ratio of hydrogen to oxygen was the same as in water, and coined the important new term 'carbohydrate'.

Liebig learnt this craft of analysis at Gay-Lussac's elbow during his time in Paris. As he said at a banquet in 1867:

I wish to speak of Gay-Lussac and Thenard. You are all familiar, gentlemen, with the great discoveries which we owe to the common efforts of these two men who were bound together in close friendship and whose works had their origin in these same friendly ties. . . . I arrived in Paris . . . a lad of nineteen, without any other recommendation than that of my desire for instruction. . . . Never shall I forget the hours passed in the laboratory of Gay-Lussac. When we had finished a successful analysis (you know without my telling you that the method and the apparatus described in our joint memoir were entirely his), he would say to me, 'Now you must dance with me just as Thenard and I always danced together when we had discovered something new.' And then we would dance.[11]

Liebig returned to Germany in 1824 as extraordinary professor at the small university of Giessen, an appointment that had required some string pulling and bending of the rules by Kastner and von Humboldt. This was because Liebig's precipitate departure from Germany two years before had left him without the necessary doctorate, and one seems to have been invented for him.

Now there was ample scope for Liebig's bull-headed enthusiasm. In only a few years he effectively founded the modern scientific department, in a model steamrollering of obdurate officials by a newly appointed academic genius. Competing professors Zimmermann and Blumhof somewhat conveniently died shortly after Liebig's arrival. While it has never been suggested even in jest that Liebig had them assassinated, it is tempting to imagine them more or less giving up the ghost in the face of the onslaught by this ferocious newcomer. Over the next two decades, in his laboratory housed in a converted barracks, Liebig and his students effectively invented organic chemistry. This was despite the fact that Liebig's own researches were very diverse and also included inorganic chemistry, instrumentation, catalysis and other things; Wöhler would fruitlessly bewail how much more Liebig would achieve if only he were more focused.

In Paris, chemical research was still an occupation for gentleman semi-amateurs, employed by the state from time to time because of their contributions to important technical problems like improving the quality of gunpowder (on which Lavoisier worked). New recruits learnt their skills by watching the professor's demonstrations, not by doing it themselves. If they proved promising, they might then be taken on piecemeal to work as his personal assistant. There was no equivalent to the modern postgraduate laboratory containing a research group working on the same or related problems. Liebig invented all this, although money was always short and his students had to pay for their own chemicals.

A description of what facilities were like before Liebig came along was given by Wöhler, describing his year working for Berzelius in 1823:

The following day I began work. I had at my disposal a platinum crucible, a pair of scales, and a wash-bottle . . . the spirit for the lamps and the oil for the bench had to be provided at one's own expense. The ordinary reagents and laboratory tools were held in common, but as prussiate of potash [potassium cyanide] for example could not be obtained in Stockholm, I had to import it from Lübeck. I was the only foreigner in the laboratory. . . . The laboratory consisted of two ordinary rooms with no water or gas pipes. One of the rooms was furnished with two ordinary tables of spruce; Berzelius had his seat at one of them, and I was seated at the other. There was also a sink and an earthenware water container where the cook, the severe Anna, daily washed the glassware . . . in the nearby kitchen, where Anna used to prepare the meals, was a little stove.[12]

These primitive facilities did not improve overnight. When they did their important early researches together, for example, Liebig and Wöhler had no thermometer capable of measuring higher than 130°C. The laboratories at Giessen were crowded and insufferably hot in summer, the students at the back climbing out of the windows from time to time to visit a beer cellar opposite.

A brief taste of the excitement of these early days of chemical research, when everything seemed possible but at the same time virtually inexplicable, is given by this quote from Wöhler. In it, he describes mixing together two compounds that he had made, one of them light yellow and the other one colourless, to produce the compound quinhydrone in 1844: 'The liquid [was] momentarily coloured a deep red, and then suddenly became full of the most beautiful green metallic prisms, which . . . are frequently an inch in length. . . . It is one of the most beautiful substances which chemistry can show . . . the nearest to it is the metallic green of the goldchafer beetle or the feathers of the humming-bird.'[13]

Chemistry can be a dangerous as well as an exciting occupation. The risks taken by nineteenth-century chemists from fire, explosion and poisoning were sometimes stupendous. The apothecary Polydore Boullay died in 1835 from the effects of a laboratory fire involving ether five years earlier. In Liebig's laboratory, the caretaker made himself a rich man by preparing the fearfully dangerous potassium metal on a large scale by distilling potassium hydroxide with red-hot charcoal and selling some of it to other laboratories. One of the high points in hazardous manipulations was reached by another famous chemist, Robert Bunsen, around 1840. In his researches on arsenic compounds, he had to breathe through long glass tubes communicating with the outside, and he partially lost the sight of an eye in an explosion.[14] Liebig introduced the fume-cupboard for all his students, and made many other advances in practical arrangements for them to carry out their work with reduced risk of an early death.

The apothecaries needed a fire lighting under them, and Liebig was the man with the bucket of red-hot coals. No longer was the Giessen pharmacy department a kind of glorified apothecary's shop that pounded up medicines and sold them to the public to pay for its upkeep. Now there were lessons in mathematics and physics and a demanding course in chemical analysis taught by Liebig himself.[15] As the years passed, the status and importance of chemistry gradually grew. Its teaching became less and less directed at these students of medicine, pharmacy and other subjects, and more and more at a group of specialists calling themselves 'Chemists'. On 1 January 1840 Liebig changed the name of his journal from *Annalen der Pharmazie* to *Annalen der Chemie und Pharmazie*. The new chemists made a takeover bid for some of the commanding heights of the apothecary profession; in 1836, for example, Wöhler became chief inspector of apothecaries' shops in the state of Hanover.

> Thanks to such a conclave [the French chemist Balard said on proposing Liebig's health at an 1867 Paris banquet], the little town of Giessen has become a flaming intellectual hearth in Germany and from its modest source has gone forth a crowd of chemists who are numbered today in all countries among the most respected masters . . . everything in his life has served the cause of scientific progress; his lessons, his examples, his incessant activity . . . even his polemic spirit, fervent yet tempered by his great kindness of heart.[16]

Analysis was the most important, and most difficult, skill that the students had to learn. Liebig improved the methods of Gay-Lussac and others, and in 1830 he designed an apparatus that was both simple in design and fairly quick to use; he completed 400 analyses in a single year. Copper oxide replaced the dangerous potassium chlorate as oxidizing agent, and both carbon dioxide and water were absorbed and collected in his ingeniously designed apparatus of glass bulbs containing potash. Thus the content by weight of both carbon and hydrogen could be fixed by weighing, which was more accurate than measuring volumes of gases. During the years after 1830, Giessen became a forcing house for the acquisition of chemical skills, especially analysis, and most of the eminent organic chemists of the next generation spent at least some time there. When a paper containing analyses was sent in to the *Annalen*, which Liebig edited, he often got his senior students to repeat and check them – an unthinkable activity for a journal editor today. The *Annalen* also had articles exposing 'Charlatanerie und Pfuscherei' (quackery and bungling). It would have been a brave man who continued life as a chemist once his name had appeared in this column.

Liebig and his students carried out many analyses of the mysterious alkaloids, including quinine, morphine and strychnine. They found that the more interesting natural products were difficult to analyse. Their molecules contain relatively large numbers of atoms that are not in simple ratios. So very precise analyses were necessary to distinguish between different closely similar formulae

differing only by very small percentages of the elements. Analyses often had to be repeated again and again by different workers ever conscious of the possibility of errors. To increase the likelihood of eliminating discrepancies, several salts or other derivatives of the natural product would be analysed one after the other.

During their most productive years Liebig was at Giessen and Wöhler had been appointed by royal decree professor of chemistry at Göttingen, 100 miles or so away but still a day's travel in the pre-railway era. An early and important fruit of their collaboration was the discovery of the phenomenon of *isomerism*.

Liebig had studied the fulminates when still a student, and, when he was in Paris working with Gay-Lussac, they had continued to fascinate him. He showed that they were the salts of an unknown parent acid, fulminic acid. At exactly the same time, Wöhler, working in Berzelius's Stockholm laboratory, was making another series of salts by the oxidation of cyanides. These he called cyanates, the salts of another (unisolated) acid, cyanic acid. They were completely different from the fulminates in their chemical properties, and were not explosive; yet, when Wöhler carried out careful analyses, he obtained exactly the same analytical results for his silver cyanate as Liebig had obtained for his silver fulminate.

In typical style, Liebig immediately accused Wöhler, whom he did not yet know, of publishing inaccurate analyses. The two met in Frankfurt, carefully compared their results and came to the conclusion that they were both right. Two substances could have the same elemental composition. Cyanic acid and fulminic acid each contained carbon, nitrogen, oxygen and hydrogen in exactly the same proportions, although the exact nature of the two acids and the relationship between them remained obscure. In 1831 Berzelius coined the term 'isomerism' to describe the phenomenon. The converse of Proust's law of constant composition is not true. A given substance with the same properties always has the same composition; but many different substances may contain the same elements in the same proportions. More and more examples of isomerism would be discovered in the following years. A formula alone was clearly not enough to describe the nature of an organic compound.

Dalton gave a tantalizing hint of where ideas about isomerism might go, in a late development of his atomic presentations that has been largely overlooked. In his absent-minded way he brushed up against, then abandoned, this important phenomenon. In his later years he drew atomic diagrams for gelatine and albumen. These are both proteins, but the term 'protein' had not then been invented, and, even when it was, Liebig thought that there was only one substance 'protein' and would have been staggered by the number we know about today.

Dalton assigned the formula C^2H^2NO to both gelatine and albumen, thus designating them as isomers. This is hopelessly inaccurate for several reasons: proteins contain thousands of atoms, not six; their composition is variable (for example, albumen from different sources differs) and albumen and gelatine are not isomers at all; their elemental composition is different. Even at the time, this must have seemed an inaccurate musing to anyone who had spent time at Giessen or in one of the other research groups now springing up. Chemistry had moved on since Dalton had first suggested his theory, and it was obvious that he had never worked in a proper laboratory. But, once again, his muse was with him, this is how he drew albumen and gelatine.[17]

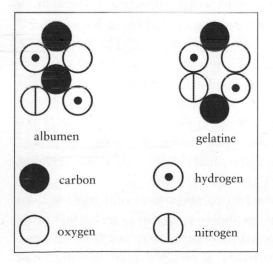

Dalton's formulae for albumen and gelatine (from Ramsay 1981).

Dalton did not offer any explanation or clarification for these pictures, or follow his idea up properly, and they were ignored by everyone else.

From about 1840, Liebig more or less switched off from the debate about the molecule. His correspondence from this period shows that he was tired of a subject that had become more and more convoluted, with no apparent resolution in sight. He may also have been tired of the disputes with Gerhardt and others (to be described), although many assaults, including Mulder's book, were still to come. Friends, including Berzelius, counselled him to stay the course. They assumed that his effective contributions to science would be over if he abandoned the subject. But they had underestimated him. By now Liebig probably lacked the mental freshness that would be needed to crack the structure problem, but, if his mind had lost a little of its flexibility, it remained adaptable and energetic, which is not quite the same thing. He became a man of affairs and business leader in the crucially important applied sciences of nutrition, public health and, especially, agriculture. In the 1860s Dumas asked him at a banquet why he had abandoned pure chemistry. Liebig replied that, once Dumas had published his substitution theory of organic chemistry (see Chapter 10), Liebig could see that the subject would require only drudges for its completion. This was a brilliantly backhanded compliment to the 'Jesuit', but of course not true. The subject was temporarily mired down. Later it would take off, but Liebig wanted to do something different.

Liebig and Wöhler, like their fulminates and cyanates, were Siamese twins, and their relationship seems genuinely to have approached the nineteenth-century ideal. Many years later, Liebig was to write: 'When we are dead and long decomposed, the ties which united us in life will always hold us together in the memory of mankind, as an infrequent example of two men who faithfully, without envy or ill-feeling, strove and disputed in the same field and yet remained throughout closely related in friendship.'[18]

Liebig became a very influential figure, one of the few hundred such who shaped the age. This has less to do with his scientific brilliance, which was patchy, than with his energy and organization. The popular view of science is that it lurches from one prescient genius – Newton, Pasteur, Einstein – to the next. This is far from the truth. Their discoveries would have been made a few years afterwards, probably piecemeal and more haphazardly, by others.

We have a tendency to hero-worship the most gifted, perhaps giving less than their due to those a little less eminent. Who among the general public has heard of Friedrich Wöhler or Eilhard Mitscherlich? What counts most is not the individual genius so much as the intellectual climate and practical environment that allows the merely talented to flourish and contribute to each other's success. Liebig was no Newton, but he did establish this productive climate throughout Europe, and so he was indeed a great scientist.

His return from Paris to found the new chemical laboratory at Giessen in 1824 was of symbolic as well as practical importance. It marked the beginning of the gradual eastward migration across the Rhine of the chemistry of Lavoisier; a migration that, despite nearly two centuries of time and two world wars, has never been reversed. Within a couple of decades, the French Chemist Boussingault was writing to his friend Dumas: 'I am never so good a Frenchman as when I am on the banks of the Rhine; it is truly shameful that an evil hole like Giessen is a focal point of science, a place where all the chemical world of Europe meets, and only comes to Paris to see our scientists as one comes to see the zoo in the botanical gardens, purely for curiosity.'[19]

This was typical of the opening shots in the war of words that would define the relations between the two countries for the rest of the century and beyond, and that would eventually be carried on by other means, chemistry playing its part.

SIX

Into the Boundless Thicket

The young suffer less from their errors than from the caution of the old.

Marquis de Vauvenargues, *Reflections and Maxims* (1746)

Paris was the centre of the chemical world when Liebig and Wöhler set out on their careers. But, whereas Liebig had made his way to that city, his subsequent friend Wöhler had with good reason headed off in a quite different direction: north. He had gone to work with the most influential chemist in Europe, the Swede Jöns Jacob Berzelius, who had an immense and controversial influence on the development of the science, which faded only in his later years.

Born in 1779, Berzelius was a youth when Lavoisier met his death, and in his mid-twenties when Liebig was born in 1803. There is something vaguely Dickensian about his early years. He was brought up in a country district of Sweden in a family of poverty-stricken clergy and lost both his parents at an early age. The young man was farmed out to various relations, all of whom had too many children of their own, and some of whom resented being saddled with him.[1] He was a family misfit, quarrelling with his cousins with whom he was brought up, until his uncle palmed him off on an apothecary as apprentice to get him out of the way. It was this that nurtured his interest in the sciences. His upbringing with its lack of love and appreciation seems to have made him a workaholic, anxious to prove himself, and in later life he fell prey to depression and lethargy. His student Wöhler, whom Berzelius in turn found 'animated, merry and dashing', described him as cheerful when working and prone to relate all manner of funny stories. 'But as soon as he was in a bad temper and had red eyes, one could expect an attack of his periodic migraine. Then he used to lock himself in, refusing to eat anything and not show up for a whole day.'[2]

Berzelius took his doctorate at Uppsala, where Scheele had once worked. His doctoral research was rather mediocre and dealt with trying to cure various diseases with electricity. These, of course, met with little success, but the electricity came to interest him more than medicine.

He is a very difficult figure to pin down. After his death most historians of science emphasized his negative attributes and came to the conclusion that chemistry might have advanced faster without him. Others have since tried to redress the balance and some are vigorously in his favour. He was the most influential thinker about chemistry of his generation, but in several ways his influence impeded the emergence of clear thought about the molecule. He was an excellent inorganic chemist, but he incorrectly tried to extend the useful ideas that he had about inorganic chemistry to the new organic chemistry, with which he was never really at home. In particular, his early electrical researches led him to the conclusion that all chemical combination can be explained in terms of electrical attraction. This is a good explanation for many inorganic compounds, but does not work (or, at least, does not work in the simplistic way that Berzelius proposed) for organic compounds.

Berzelius certainly felt that organic chemistry was a subject that ought to be probed, and did so with enthusiasm, at least initially. In 1814 he wrote to the London-based Swiss chemist Marcet (husband of Jane Haldimand, who wrote the *Conversations on Chemistry*): 'I am still going on with my studies of . . . organic nature. It is extraordinarily difficult and the difficulties increase as I try to study labile substances which break down as one is in the process of isolating them from other substances . . . this study is the most difficult I have ever worked on.'[3] He did not lack confidence, however: 'You will see how clear chemistry will become and what a fine foundation its theory will acquire through these studies.' He made some important early contributions to the new science. He was one of the first to recognize the need to define what was meant by a pure compound, and showed that many of the natural products of the apothecary were in fact mixtures. Although organic matter might be less 'well behaved' than inorganic, the need to identify what were the pure compounds remained. But later he became disillusioned with the subject. His incorrect theoretical ideas were tying him in knots.

More reprehensibly, he had a very dogmatic approach that, like Liebig's argumentativeness but in a different way, discouraged free debate. It is one thing to be wrong, but another to be so fixed in your views that others become inhibited, and that was certainly the case when Berzelius was virtually judge and jury for chemical ideas. According to Melhado, Berzelius can be classified as a 'synthetic' thinker.[4] This is someone who is always on the lookout for new connections, a philosophical approach that was certainly in accord with the predominant intellectual climate in Europe during the Biedermeier period. Berzelius advanced many new hypotheses and often drew analogies between what appeared to be unrelated phenomena, sometimes incorrectly and confusingly. But, at the same time, he was of a conservative turn of mind. Like the majority of people but to a greater degree than most, he could accept new ideas much more readily when they came from his own brain rather than from others. He said that such a turbulent science as chemistry should not throw the older ideas needlessly overboard. 'Chemists should not advance theories without considering how they would affect chemistry as a whole,' he said once, a stricture that, if one were unkind, one could say Berzelius never applied to himself. His tone on this occasion was 'very superior'.[5]

He could also be dogmatic about subjects that he had hardly studied – for example, physiology. 'The small animals in semen shaped like an egg with a short tail at the end may sometimes be lacking without reduction in its fertility' was one of his pronouncements. Another concerned green vegetables, such as asparagus and cabbage: 'Some of them contain little, and most of them no nourishment; they are used by the poor to give them what we call a full belly.'[6]

Because of the excellence of his early experimental work, Berzelius gained great prestige. He totally dominated chemistry in Sweden and far afield, and never had any difficulty in getting his ideas published. The new generation, such as Wöhler, tended to defer to his views and to some extent leave him master of the field of theoretical speculation while they got on with their experimental work. Later on, Berzelius's influence faded, but Wöhler, like Liebig, had, on his own admission, by then lost interest in theoretical speculations; he had spent too much time in the 'boundless thicket of organic chemistry', which had partly sprung up as a result of plantings by Berzelius.

The most important way in which Berzelius's ideas were counter-productive concerned his views on the importance of electricity in chemistry.

In the eighteenth century, electricity had been extensively studied, although the means for generating it remained electrostatic. It was realized that some substances conducted electricity while others did not. Positive and negative electricity were discovered, and it was shown that electrical attraction and repulsion obeyed an inverse square law, like Newton's gravity. By 1767 there was already enough known about electricity for Joseph Priestley to consider it worth writing a history of it.[7] The phenomenon continued to fascinate the general public. Demonstrators of electricity toured Europe, culminating in an occasion when the French priest Jean-Antoine Nollet persuaded 700 monks to link hands and receive an electric shock when the hands of those at the ends of the chain were connected to the terminals of a Leiden jar. Most of them said that they had gained a brief taste of what hell was like.

Chemistry and electricity were clearly intimately related, but in unknown ways; the voltaic pile or battery converted chemical energy into electrical energy, and the experiments on electrolysis by Davy, Faraday and others converted it back again. Not only that, but Galvani's experiments on the nerves and muscles of frogs showed that the functioning of the organism also depended on this mysterious fluid, which somehow seemed related to the life force itself. The powerful poison strychnine caused convulsions in the muscles that were very like those induced by a powerful electric current, and that, if sufficiently powerful, led to death. It was widely thought that anyone touching a strychnine-poisoned human being or animal received an electric shock.

From this it was only a short step to try to resuscitate dead people by passing electricity through them. In 1818 Dr Ure in Glasgow tried to resurrect the corpse of a murderer using galvanic batteries. 'Every muscle in his countenance was simultaneously thrown into fearful action; rage, horror, despair, anguish and ghastly smiles united in their hideous expression in the murderer's face . . . at this point, several of the spectators were forced to leave the apartment from terror or sickness, and one gentleman fainted.'[8]

An even more profound link between electricity and the life force was described a few years later in 1837 in some notorious experiments by Andrew Crosse. He claimed that operation of a battery in his laboratory caused the spontaneous generation of flies in the electrolyte. While even at that time almost any responsible scientist would have written off his claim as rubbish, the very fact that he could get it published in the *American Journal of Science* showed that the lunatic stratum was not far from the surface where anything involving electricity was involved.[9]

An important tenet of the *Naturphilosophie* was a belief in dualism. Some German philosophers, including Kant, thought that matter was the result of an equilibrium between an attractive and a repulsive force, known as the 'dynamic' picture. They believed that, if the attractive force were to prevail, the universe would collapse to a point – an early prediction of black holes. This dualistic philosophical framework fostered the development of scientific ideas in which electricity played a central role. The dominance of these ideas was temporary. By about 1840 *Naturphilosophie* was going out of fashion and most practical scientists felt that such ideas had little to offer. By this date Liebig was attacking the intellectual climate in Prussia, claiming that there was no chemical laboratory worth the name in the entire country, and blaming this on the prevalence of woolly philosophical teaching. He referred to the 'False Goddess' of the *Naturphilosophie* as a 'Black Death'. But, in Berzelius's hands thirty years before, it had taken organic chemistry down a blind alley.

In 1812 Berzelius had wanted to visit France, but war had broken out between France and Sweden, so he went to England instead. Here he met Davy, virtually a contemporary of his. Davy had carried out important research on generating electricity by means of various kinds of batteries, and at one point had gone so far as to state that electricity was a wholly chemical phenomenon, although by 1806 he had abandoned this idea.

Like others before him (including Lavoisier), Davy had postulated that the 'fixed alkalis', soda and potash, might not be elements at all but compounds of yet undiscovered metals. He thought it might be possible to isolate these using power from the bigger and bigger voltaic piles, or batteries, now being made.

On 6 October 1807 Davy carried out one of the most spectacular experiments in the history of science, which he referred to in his notebook as 'Capital Experiment proving the decompn of Potash'. He placed a piece of pure potash (potassium hydroxide) on a platinum disc connected to the negative side of a 250-plate battery, and touched it with a platinum wire connected to the positive terminal. The potash began to melt and give off gas at the upper terminal; at the lower plate, small globules of shiny metal looking for all the world like mercury began to appear, some of which immediately burnt with a bright explosion, while others became tarnished in the air. When thrown into water, the globules of potassium took fire and burnt with a purple flame. According to his assistant, Davy 'bounded about the room in ecstatic delight, and some little time was required for him to compose himself sufficiently to continue the experiment'.[10]

This and similar experiments were highly influential. They confirmed the fact that matter, if not actually composed of electricity, was at the very least intimately related to it in some as yet unknown fashion. Berzelius, who had obtained his MD degree in 1802 with his thesis on the medical applications of galvanism, was one of those who did most to emphasize its importance. He continued to exchange letters with the famous chemist in London, and, as a result of their correspondence, Davy went on to prepare the metallic elements barium, calcium, strontium, magnesium and lithium for the first time. Berzelius's thinking on the subject of chemical constitution thus became centred on ideas of electrical dualism: positive and negatively charged aspects of the substance.

According to his theories, it was possible to sort the elements into electropositive ones, such as potassium, and electronegative ones, such as chlorine or oxygen. Chemical compounds resulted from a positive part coming together with a negative part. In the simplest case, the two parts were elements, say potassium and oxygen. But the elements differed in the degree of their electropositive or electronegative nature. Since potassium was more electropositive than oxygen was electronegative, the resulting oxide, a 'first-order compound', would have a residual electropositive nature that would enable it to participate in further reactions to produce compounds of the second order, and so on.

According to Berzelius's dualistic formulae, based on electrochemical ideas, every substance was composed of an electropositive and an electronegative part. Each part may consist either of a single atom (as in sodium chloride) or a complex radical, in its turn composed of a group of atoms. Brackets were later used to represent the holding-together of the two electrical parts. For example, this typical bracketed formula, dating from the late 1830s, is how he represented ethyl acetate.

$$\left.\begin{matrix} C^2H^6 & O \\ C^2H^6 \\ O^2 \end{matrix}\right\} O$$

The difficulties raised by these formulae are obvious. If the only force allowed to glue atoms together is electrical, how are the radicals themselves held together internally if not by electricity? If this is the case, how does the way in which they are held together internally differ from the way in which the two radicals are held together? The logical extension is that some other force is involved. Could this be the vital force?

Dalton had postulated that, as well as binary combinations AB, A_2B, AB_2, combinations such as A_2B_2 and A_2B_3 might exist, although he could not see exactly how A_2B_2 might differ from AB plus AB. This is indeed difficult using Berzelius's ideas. If A_2B_2 is not the same as AB, then the two As and the two Bs must be held together in some way. If the only kind of bonding is electrical, this is impossible. Berzelius in 1815 poured scorn on the idea. Dalton had presciently if tentatively suggested that olefiant gas, or ethylene, might be C_2H_2 (it is actually C_2H_4). Berzelius called this a 'mere dream'. Such views, he said, threaten the solidity of the whole theoretical system, and his own views were based on a comprehensive survey of the composition of compounds.[11]

Dualism would in the end turn out to be an unproductive line of speculation. It was based on defective physics and was largely ignored by physicists, though most chemists took it up enthusiastically. It was not too bad a working hypothesis for describing the way simple inorganic salts were held together, but when it came to organic compounds it was seriously wanting.

To understand why, we need to jump forward about 100 years to a description of the structure of the atom and of the chemical bond, as worked out by chemists and physicists in the opening decades of the twentieth century. It would be possible to update this description to the present time and arrive at something more sophisticated, but the description given here is sufficient to explain the phenomena that so tantalized chemists until the 1860s.

Atoms consist of a very small, positively charged nucleus containing one or more protons, each one carrying a positive electrical charge. The number of protons determines the element. This nucleus is surrounded by a cloud of much lighter negatively charged electrons. In the neutral element the number of electrons equals the number of protons, so that the overall electrical charge is zero. The number of protons/electrons determines the element; hydrogen, the lightest element, has just one of each.

The chemical behaviour of each element is dominated by the orbiting electrons. Early versions of the atomic theory saw these as negatively charged particles circling the nucleus like tiny planets round the sun. Later and more sophisticated versions of the theory say that they cannot be said to orbit the nucleus in such a straightforward way, because they are partly particles and partly waves. But this does not affect the argument. We will continue to think of them as particles.

The electrons are in 'shells'. Atoms are happiest when they have attained a closed shell. This is most frequently of eight electrons, but the innermost shell is full when it contains only two. The reasons for these numbers lie deep in quantum theory. The resulting fully satisfied electronic structure with an outer shell of two or eight electrons corresponds to that of the unreactive 'noble' gases, helium, neon, argon, krypton and xenon, which are present in the atmosphere, but, because of their inertness, were not discovered until 1894.

By the early nineteenth century, it was realized that certain of the elements seemed to be closely related in their properties. One such group was christened the *halogens*, or salt-formers. The most ubiquitous of these is chlorine, a poisonous yellow-green gas in its elemental form but one that is all around us in the form of ordinary salt (sodium chloride) and a host of other everyday compounds. Initially there was confusion about whether chlorine was in fact an element, or whether it contained oxygen, as Lavoisier thought.[12] Its status as an element was confirmed by Davy and others around 1810 and within a very short space of time it had been joined by two other halogens, the violet crystalline iodine that Davy himself discovered, and the choking red liquid bromine. Like chlorine, bromine and iodine are reactive and are commoner as salts than in the elemental form.

The halogens are the family of elements in which the outer shell is of seven electrons; their chemistry is dominated by the tendency to gain the additional electron required for the inert or noble gas structure. Conversely, the alkali metals such as sodium and potassium discovered by Davy have an outer shell of one electron that is very readily donated. When sodium has reacted with chlorine, the number of electrons on the sodium no longer balances the number of protons in the nucleus; it has become a sodium *ion* carrying a positive charge, while the chlorine has become a negatively charged chloride ion. The resulting crystal of salt is an array of positively charged sodium ions alternating in a giant three-dimensional lattice with negative chloride ions. The input of large amounts of electrical energy, as in Davy's experiment, is sufficient to neutralize the charges and generate the reactive free elements from the stable salt. (He initially electrolysed potassium hydroxide rather than sodium chloride, but the principle is the same.)

But elements with three, five or especially four electrons in the outer shell are less likely to be able to gain or lose the requisite number of electrons to attain the noble gas structure. To pull away or push in four electrons generates a large electrostatic charge that is unfavourable. These elements, of which carbon with its four outer electrons is for us the most significant, have to bond in other ways, which, although still based on electrons, are not based on ion formation and therefore are not 'electrical' in the straightforward sense. They form bonds by *sharing* pairs of electrons with other atoms of the same or different elements. If the two elements thus bonded together have similar electronegativity (electron-attracting power), then the resulting bond is electrically neutral. It also has direction in space.

So attempts to explain the constitution of organic compounds on an electrical basis *in Berzelius's simple way* were doomed to eventual failure. It is mysterious why Berzelius, with increasing desperation as the years passed, tried to shoehorn organic compounds into his electrical bonding theory if at the same time he thought they were uniquely full of vital force and not subject to the normal laws of chemistry.

Ideas of chemical structure dominated by charged bodies could not deal with the most important feature of bonding in organic compounds, the phenomenon of *catenation*, or the ability of atoms to form chains and rings by means of stable chemical bonds between themselves, especially in the case of carbon. This is a vital phenomenon explaining bonding in organic compounds. It would be another half-century before the idea of catenation emerged properly.

In the opinion of some, Berzelius's ideas have received posthumous validation from what we now know about the details of the way that molecules are held together. But modern ideas of electrons and orbitals are several intellectual generations away from Berzelius's views, and it is always suspect to credit someone with foresight when their ideas are counterproductive in the shorter term, which Berzelius's certainly were.

Berzelius also recognized at an early date the need for a better system of notation for chemistry. The development of a science is intimately related to the symbols used to represent it. In the alchemists' system used for chemistry until the end of the eighteenth century, each substance had its own symbol, varying between different authors. Lavoisier, for example, used ▽ for water, a symbol that could trace its ancestry back to the thirteenth century. These symbols did little more than replace the frequently obscure alchemists' names by equally obscure symbols, although by about 1775 chemists were trying to use them in diagrams to represent chemical reactions.

The possibility of something better was ushered in by Lavoisier's list of elements, and two systems for representing them came into existence.[13] One was that of Dalton, who in effect adopted the pictographic system of the alchemists, with the crucial difference that the symbols now represented elements not individual compounds. Thus Dalton's symbols for carbon, hydrogen and nitrogen were as follows:

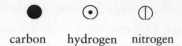

carbon   hydrogen   nitrogen

Berzelius invented another notation in 1811. This is the familiar one based on alphabetical symbols for the elements such as C for carbon, O for oxygen and Pb (Latin, *plumbum*) for lead, together with subscript (originally superscript) numbers showing the number of atoms combining.

This symbolism has a number of plus points. It is easy to write, remember and typeset. It enabled chemistry to be based on a more mathematical footing and helped clarify thinking about chemical reactions. From the early nineteenth century onwards, as a result of Berzelius's notation, equations could be written using an equals sign to emphasize that the sum of the products on each side of the equation should balance. As an example, suppose we write the reaction between sodium metal and water to form sodium hydroxide, using modern formulae, like this:

$$Na + H_2O = NaOH$$

It is clear that some of the hydrogen has 'gone missing'; the equation does not balance. We have forgotten the hydrogen gas produced, and, if we had not noticed it bubbling off, we might go back to look for it. The correct equation is

$$2Na + 2H_2O = 2NaOH + H_2$$

which balances.

The Berzelius notation thus helped the development of chemistry and has stood the test of time. It is familiar to every schoolchild. And yet somehow, by using it, the mental connection between the symbolism and the physical reality is weakened.[14] This comes about in at least three ways.

First, consider the element symbols themselves. Which of the two notations, 'C' or '●' best conveys the fact that the atom is a physical entity, a 'little thing'? '●' surely resembles a little picture of one of Dalton's wooden ball models, and implies that atoms exist and that they have size and position in space. 'C' does not.

Secondly, how about the region between the 'balls' or what we, with hindsight, would call the 'bonds'? If Dalton's symbols had continued in use, or even if Berzelius had written water as H–O–H, someone might eventually have enquired what might be represented by the lines between the atoms, or to ask why write it H–O–H and not H–H–O, for example. But the formula H_2O says nothing about possible structure.

The third point is more subtle. When a chemist wrote down 'Na', what exactly did this represent? Did it mean *some* (unspecified amount of) *sodium*, *one atom of sodium*, or *one equivalent of sodium* – that is, the exact amount of sodium necessary to take part in the reaction with an equivalent weight of water, and thus balance the equation?

These points were made much later in 1872 by the English chemist C.R.A. Wright (the discoverer of heroin) in a letter to Sir Benjamin Brodie.[15] 'It seems to me', he wrote, 'that the formula H_2O and the pictorial symbol ⊙–O–⊙ represent two different sets of ideas . . . the former just the facts that 2 vols. of steam contain 2 of hydrogen and one of oxygen . . . the latter represents the theory of the existence of finite indivisible portions of matter and a notion as to their mode of union or connection.' (Wright was a sceptic about atoms.) It is remarkable that, sixty years after Berzelius introduced his notation, it still remained possible for a chemist to comment on the unresolved intellectual fog surrounding the whole business.

Because the Berzelius system prevailed in the end, it is easy to assume that it came after Dalton's, and supplanted it. But they were more or less synchronous. Dalton lectured on his atomic theory until at least 1837, and his symbol system persisted among many chemists in England until it gradually died out at about the time of his death. Not surprisingly, Dalton in a letter to Thomas Graham called Berzelius's symbols 'horrifying' and compared them with Hebrew letters. Dalton's system died out because it was less convenient, but also because at the time of his death there was still no firm evidence for the existence of his atoms.

It is true that, in the first half of the nineteenth century, a sizeable proportion of chemists did not believe in the atomic theory. But Berzelius was not among them. He was a firm believer in atoms. One of his main objectives became to establish the truth of the atomic theory through the application of electrical principles, although characteristically, and counterproductively, he invented his own take on the atomic theory, his corpuscular theory, which served to muddy the water. Paradoxically, the symbolism that he introduced may have inhibited thought about the very problems that he began by being so confident of solving.

Chemistry was at a stage in its development as a science in which a proper balance had to be struck between experimental work and theoretical speculation. More importantly, for speculation to be productive, there needed to be a climate of free, open and unintimidating exchange of ideas. It cannot be said that this climate always predominated. In the early decades of the century, Berzelius exerted a stranglehold. As his influence gradually waned, various shifting alliances developed between Liebig and other German chemists, and those in other parts of Europe, especially France. Liebig is partly to blame, first for abandoning most of the theoretical part to Berzelius, then later for being so combative in his published work that it would have been a brave person who would venture into the argument. Holmes sees this as a natural consequence of the helter-skelter development of a young science.[16] The protagonists, he says, were 'being swept along in a moving investigative stream that none of them could control', but this is being generous to them. Despite his brilliant experimental work, Berzelius's contributions had many negative aspects, not the least his encouragement of such an ill-tempered debate. His contributions to the central dilemma of atomic weight were problematical.

SEVEN

May We See the Trees?

*Chemistry is at heart only a kind of arithmetical exercise, that
sometimes pleases because it makes sense and the sums add up;
but in the end, its purpose is no more than making a good
bootpolish, or cooking meat to make it taste better.*
Justus von Liebig, in a letter to Wöhler, 1 May 1832, in Schwarz
(1958)

The central problem with which chemists wrestled for the best
part of the nineteenth century was a deceptively simple one.
They needed accurate atomic weights of the elements, the ratio by
which the atom of one element is heavier than that of another.
Without these, it was impossible to assign accurate *formulae* to the
compounds, and all speculation about whether or not the molecules
had meaningful *structures* was doomed to disappointment.

Atoms, if they existed at all, were far too small either to see or to
measure directly in any way. Chemists could not directly measure
atomic weights; they could measure only *equivalent weights*. These
are the proportions by weight of one element combining with
another. Eight grams, or grains, or tons (to the nearest whole
number) of oxygen combines with one gram, or grain, or ton of
hydrogen to form water. The equivalent weight of oxygen in water is
8. (It was usual, though not universal, to put the weight of the
lightest element, hydrogen, equal to 1. Other systems were also
used; for example, Gay-Lussac used as his baseline oxygen = 10.
This did not affect the main argument, but it did add another layer
of complexity.)

Berzelius, a much better experimenter than Dalton and a strong
believer in the atomic theory, set out with a main aim to find out the
atomic weights of all the known elements by meticulously measuring

their equivalent, or combining, weights. By 1818 he had published a table containing the percentage weight compositions of nearly all of the 2,000 chemical compounds then known. On this he based what he thought were the atomic weights of forty-five of the forty-nine known elements. Six of the determinations were by his pupils and the other thirty-nine by Berzelius himself. It was a remarkable achievement.

But equivalent weights do not give the atomic weights. Certain assumptions have to be made to go from one to the other. Assume that you were a chemist who firmly believed in Dalton's theory. According to your view, when oxygen reacts with hydrogen to form water, a large but unknown number, n, of oxygen atoms were combining with hydrogen atoms. But was the number of hydrogen atoms n, $2n$, $\frac{1}{2}n$ or some other number? Was the formula of water HO, H_2O or HO_2? There was no direct way of knowing. This problem of getting from the equivalent weights, which were experimental facts but could not be properly explained, to the atomic weights, which if properly known would explain everything but which could not be directly determined, would dog chemistry for half a century and more.

Imagine that you are with von Humboldt and Bonpland on an expedition through South America. You encounter a people that is little known in Europe, and you have been asked by the French Academy to find out as much as you can about their social habits. Of the few explorers who have met them before, some assert that they are monogamous, some polygamous and some polyandrous (that is, one woman taking several husbands). There are even rumours that they practise completely novel arrangements, such as two women having three husbands between them. You have no idea how many people there are in the tribe; it might be 100, it might be 50,000. The Head Man refuses to let you count them, or to divulge the secret of their marital arrangements. He does let slip, however, the fact that, whenever someone gets married, by custom they plant a number of ceroxylon trees, so many for a man and so many for a woman. 'May we see the trees?', Bonpland asks. The chief declines. The only fact he will divulge is that the men's plantation is eight times the size of the women's. 'There you are,' exclaims Dr Dalton, who happens to be along (it is an unusual expedition). 'Nothing could be clearer. They are monogamous (which is in accordance with God's law), and every man plants eight trees when he gets married, and every woman one.'

Bonpland disagrees. 'I often see a man crossing the village square with two women. I think they have two wives, and plant sixteen trees.' Humboldt, however, at great personal risk, peeped into the communal bathhouse yesterday and counted twelve men and only six women. So the men must plant four trees . . . substitute weights for trees, and you have a rough analogy of the chemists' problem. How to resolve it?

At first the only answer was to carry out a large number of experiments and comparisons to try to determine which ratios made the most sense. Dalton used Occam's razor, the principle invented by the fourteenth-century English friar William of Occam: when there are two or more competing theories that make the same prediction, you should choose the simplest. (Dalton was English, after all; and so was William of Occam. To go for the simplest explanation is an English trait. It often works, but sometimes does not. It is difficult to imagine a French monk coming up with such a principle.)

Dalton said that, when two elements combined to form only one binary compound, the ratio was always 1:1, and tenaciously stuck to his HO formula for water when evidence mounted up to show that it was wrong. In other cases, Dalton 'patched up' his theory without real evidence. By 1814 others were already criticizing him for ignoring Occam's razor in guessing some formulae. John Bostock, exasperated, wrote:

> When bodies unite only in one proportion, whence do we learn that the combination must be binary? Why is it not probable, that water is formed of two atoms of oxigen and one of hidrogen, of two atoms of hidrogen and one of oxigen, or in that of any assignable number of atoms of hidrogen and oxigen? I do not perceive that Mr Dalton has given any reason in support of this binary combination, in preference to all the rest, and I am unable to conjecture what reason can be urged in its favour.[1]

The formulae of various other oxides were also guessed. If an element only formed one oxide, or if one oxide was more stable or 'commoner' than the others, they used Occam's razor to give it the formula XO, and called it the 'Protoxide', not always correctly.

Dalton defended his choice for water by insisting the combination with the smallest number of atoms, such as HO, would give the least amount of repulsion between like atoms. But, even leaving aside for a moment that this is wrong in fact for water, which is H_2O not HO, his explanation cannot even give a decision in other cases. One example was copper. This forms a black oxide and a red oxide that contains half the amount of oxygen. The black oxide could be CuO and the red oxide Cu_2O; or the black oxide CuO_2 and the red one CuO. In the first pair, the red oxide contains two copper atoms, and, in the second pair, the black oxide contains two oxygen atoms. So Dalton's explanation based on repulsion is useless. Because the black oxide decomposes fairly readily into copper and oxygen, he assumed that it is CuO – as it happens, correctly; but his reasoning was not valid.[2] Incorrect choices by Dalton, Berzelius and others led through a whole network of interrelations to many incorrect formulae. In the vital case of carbon, although Berzelius started off with the correct atomic weight of 12, before long this was amended to 6, a figure that many chemists used until the 1850s, with disastrous consequences.

The method of building up a network of chemical inter-relationships and hence a mass of equivalent weights, so assiduously carried out by Berzelius and others, needed at least one other method to cross-check it. Before long, several other methods were proposed. But none of these was known to be valid. Each one was a hypothesis, not an established law. And each one gave results that sometimes agreed with the equivalent weight method, and with each other, and sometimes disagreed.

The first method was to measure the volumes of gases that combined in chemical reactions. This led from the work of Gay-Lussac, who in 1808 found that, when gases combine, they do so in simple ratios by volume. One volume of hydrogen combines with one volume of chlorine, for example, while one volume of oxygen combines with two of hydrogen. Dalton, irritatingly, did not at first accept Gay-Lussac's law, although it would seem on first sight to provide an excellent proof of his atomic theory. He said that the closest approach to 2 that he could get for hydrogen and oxygen in his experiments was 1.97. The concept of experimental error never seems to have appealed to him, especially when it implied that he was a rather poor experimenter compared with the Parisian chemists.

On the basis of Gay-Lussac's law, Davy advanced the correct formula H_2O for water, but Faraday argued against him and in favour of HO. Since water could be electrolysed – that is, broken down by electricity into oxygen and hydrogen – he said, it must be HO, because he incorrectly thought that all electrolytes must be 1:1 compounds. This shows the difficulty that chemists had in choosing between different suggestions for arriving at the formulae. H_2O was generally accepted after about 1819, when the influential Berzelius adopted it. He accepted that there is no reason to suppose that a litre of oxygen and a litre of hydrogen should not contain the same number of atoms. This is how he saw the combination of the two gases. (In these diagrams, an arbitrary six atoms are shown in each box of equal volume. The true number in any given visible volume of gas was, of course, unknown but very large.)

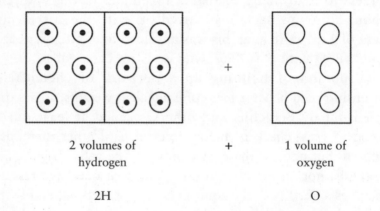

2 volumes of   +   1 volume of
hydrogen   oxygen

2H   O

But there was a huge difficulty. Gay-Lussac stated his law to refer to the reactants only. If these diagrams involving the atomic theory are correct, then the volumes of the *products*, if they are gaseous, should surely also fall into line. But they do not. The volume of steam produced from hydrogen and oxygen is twice that expected. An even more disturbing example is the reaction of nitrogen and oxygen to form nitric oxide. One volume of nitrogen combines with one of oxygen to produce *two* volumes of product. If the extension of Gay-Lussac's law to products as well as reactants is valid, this can be true only if the 'compound atom' of nitric oxide contains *half* an atom of oxygen and *half* an atom of nitrogen!

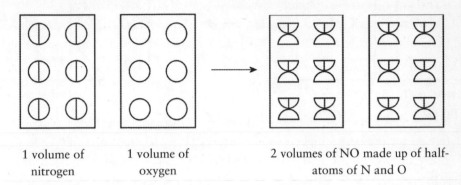

1 volume of nitrogen	1 volume of oxygen	2 volumes of NO made up of half-atoms of N and O

Combination of one volume of nitrogen with one volume of oxygen to produce two volumes of nitric oxide showing that, if Gay-Lussac's law applies to both reactants and products, 'half-atoms' were thought to be necessary.

But the atom is, *by definition*, the smallest indivisible particle of matter; *therefore*, the reasoning must be faulty. *Therefore*, Gay-Lussac's law cannot apply to the products of reactions. *Therefore*, it may not even be true for the starting materials and *therefore* the whole of Dalton's atomic theory may be the ill-conceived dream of an absent-minded schoolteacher from the Cumbrian mists.

Berzelius, a convinced atomist, sidestepped the nitric oxide difficulty by proposing that only elements and not compounds obey Gay-Lussac's law. There is no reason to suppose that this should be true. The correct explanation is that the molecules of hydrogen, oxygen and nitrogen are diatomic – that is, they are each made up of two atoms. The atom and the molecule are not the same thing. Here is the correct representation of the two reactions.

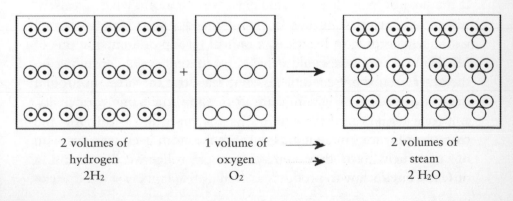

2 volumes of hydrogen $2H_2$	1 volume of oxygen O_2	2 volumes of steam $2 H_2O$

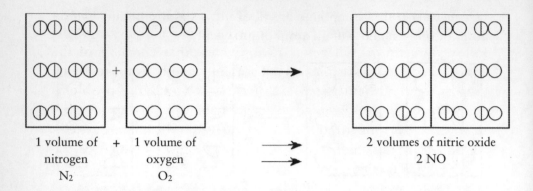

<table>
<tr><td>1 volume of
nitrogen
N_2</td><td>+</td><td>1 volume of
oxygen
O_2</td><td>⟶
⟶</td><td>2 volumes of nitric oxide
2 NO</td></tr>
</table>

Berzelius was excruciatingly close to the correct explanation. But he had gone off on a diversion that distracted him from the main argument. Unable, seemingly, to adopt Dalton's theories at face value because they were not his own invention, he had developed his own 'corpuscular' and 'volume' theories as alternatives. He was currently engaged in arguments with Dalton about the relative merits of their theories and was trying to fit the facts into his own theoretical framework, not into Dalton's. The situation was confused, and the confusion was largely of Berzelius's making. One critic wrote: 'How the substitution of the word *volume* for *atom* simplifies the atomic theory . . . is beyond my comprehension'.[3]

The bizarre explanation that the atoms are divisible was resurrected later in different ways by both Dumas and Laurent. Dumas said that, in the nitric oxide reaction, the entities combining were indeed 'half-atoms'. This rightly brought scorn from Berzelius. If the smallest particle that could exist was an atom, what, precisely, was a half-atom? Laurent later advanced a more sophisticated variant. According to Berzelius, a radical (group of atoms) in one of his dualistic formulae could behave in different ways depending on the other groups present and on what reactions the substance found itself involved in. Might not an atom show the same propensity, changing its nature depending on its environment? If this was the case, the only explanation must be that the atom itself was made up of constituent parts that could rearrange to behave differently in different circumstances.

Depending on one's point of view, this was either a lunatic idea or a breathtakingly prescient insight into the subatomic physics of the following century. What was clear to the other chemists of the 1840s, however, was that it was not a useful hypothesis for resolving their present difficulties. Most of them were having problems enough coping with the implication of Dalton's atomic theory. The suggestion that his atoms were not really the ultimate particles seemed to throw chemistry back into the arcane world of the metaphysical; no thank you.

So, measuring the volumes of combining gases and their products did not seem to hold the key to atomic weights. Three other physical methods were proposed that it was thought might give objective answers.

The first of these was the measurement of the densities of various gases, including the vapours of substances that were liquid at room temperature. Dumas developed a special apparatus for doing this and made many accurate measurements. If gas A was twice the density of gas B, this implied that its atomic weight was greater by this factor of two. The method could be used for compounds as well as elements. Dumas at first thought that this would hold the key to all atomic weights. The method relies on the assumption that the atoms all occupy the same volume, as Gay-Lussac's law would seem to imply.

Leaving aside the obvious difficulty that many elements and compounds could not be studied in the gas phase because they were not volatile enough, the method threw up strange anomalies. The results given by the common elements oxygen, hydrogen, nitrogen and the halogens (such as chlorine) were self-consistent. The reason is that all these gases are diatomic – that is, the molecule is composed of two atoms: O–O or O_2, H_2, N_2, Cl_2. But, even among the elements, mercury gave a figure half, and sulphur a figure three times, the values expected from the other methods. The explanation is a simple one. Mercury vapour is monatomic, Hg, while sulphur vapour consists of rings of linked sulphur atoms, S_5, S_6, S_7 and so on, with an average size roughly equivalent to hexaatomic sulphur, S_6. The concept that was missing was that even elements might have molecules composed of more than one atom.

A second method was proposed, or rather invented, by French chemists Pierre Dulong and Alexis Petit. Dulong, who lost two fingers and an eye in a laboratory accident involving nitrogen trichloride, and Petit, who died at the age of twenty-nine, were both brilliant chemists, and in 1819 stated their law that 'the atoms of all simple bodies [elements] have exactly the same capacity for heat'. This law, or rather hypothesis, was based on only a few experimental observations at the time. Measuring the specific heat, or heat capacity, of an element and multiplying it by the atomic weight give the same figure for each element, to within a few per cent. More to the point, by measuring the specific heat of a new element and dividing it by the standard number, they could obtain its atomic weight. Their method was not particularly accurate, but accuracy was not necessary. It could be used to give a choice between, for example, approximately 23 and approximately 46 for sodium. But it was not known to what extent their 'law' was true. It is essentially correct, but there are complications. Most importantly, the specific heat of carbon (diamond or graphite) is inexplicably low. When heated to high temperatures, its value approaches the normal one, but this was not found out until 1871. When Dulong and Petit tried to change the atomic weights of sulphur and silver, Berzelius accepted the former change, but rejected the latter one. Berzelius remained the final court of appeal.

The third method came from the work of Berzelius's associate, the German chemist Eilhard Mitscherlich, who in the same year, 1819, discovered the phenomenon of isomorphism. Two substances sometimes have the same crystalline form and could form mixed crystals together – for example, sodium sulphate and silver sulphate.

Berzelius seized on this method triumphantly. If isomorphism was possible, the ratio of the weights of the two elements replacing each other should be the ratio of their atomic weights. The discovery was 'the most important made since the doctrine of chemical proportions. . . . [Mineralogy] hastens now with wings towards its development and leaves the mere natural historians behind', he wrote.[4]

But again, the high-flown rhetoric was premature. Like Dulong and Petit's law, isomorphism had only limited applicability. More disturbingly, it began to emerge that some pairs of salts gave results inconsistent with other pairs of salts featuring the same elements. How to distinguish the good analogies from the bad? Berzelius thought that sodium sulphate/silver sulphate was a valid example of isomorphism and that it confirmed his atomic weight assignments of the two metals. He was both right and wrong; it *is* a valid example of isomorphism, between Na_2SO_4 and Ag_2SO_4, but it does *not* confirm his atomic weights; it merely indicates that, *if* sodium is 46, *then* silver is 216, but in fact he had got both figures wrong. What a mare's nest!

Thus the several methods of atomic weight determination gave results that were often in agreement, but almost equally often in disagreement. Yet, as Berzelius noted, when they were in disagreement, the anomaly was always close to a whole number; the experimental value by one method was twice, or three times, or half, what the other methods predicted. Did not this in itself point to an underlying regularity presumably related to the atomic hypothesis?

Berzelius and others published successive tables of atomic weights, taking into account new and more accurate measurements, and recording the atomic weights of newly discovered elements. But these attempts were premature. Bradley refers to the 'astonishing complexity and treacherous difficulty' of the problem.[5] The inaccuracies were of two kinds. The first resulted from inaccuracies in experiments. When dealing with natural products that might contain thirty or forty carbon atoms, it made a difference whether the atomic weight of carbon was, say, 12.1 or 12.2. These inaccuracies gradually diminished with ever more meticulous experiments. For example, Dumas and his co-worker Stas made a series of thirteen very accurate determinations of the combining weights of carbon (graphite or diamond) and oxygen, and came up with the most accurate figure for carbon's atomic weight yet found. But the real problem came from the 'order-of-magnitude' uncertainties. Was the atomic weight of carbon, to the nearest whole number, actually 12 or 6?

On the basis of isomorphism experiments, Berzelius made several changes to his atomic weights table. Then Dumas made a number of other changes. Then Gerhardt reviewed the whole subject and made far-reaching further changes in 1846. But by this time there were so many values around that many groups of chemists were using obsolete values. The vital atomic weight of carbon went from 12 (Berzelius) to 6 and eventually back to 12 again.

These atomic weight uncertainties adversely affected every attempt to assign a molecular formula on the basis of analytical results. These could not be put right by more accurate analyses. To give an example, when Liebig and Wöhler analysed the important substance benzoic acid in 1832, they combusted 523 milligrams of the acid in Liebig's apparatus and obtained 1,308 milligrams of carbon dioxide and 238 milligrams of water. By a simple arithmetical exercise involving dividing by the presumed atomic weights of the three elements carbon, hydrogen and oxygen, this gave a formula of $C^7H^6O^2$, the correct value, or a multiple thereof. Any multiple of $C^7H^6O^2$, the 'empirical formula', would give the same analytical figures. They then made the silver salt of the acid and analysed that. But, because they were using Berzelius's incorrect figure of 216.6 for the atomic weight of silver (twice the correct value), they opted for the incorrect formula $C^{14}H^{12}O^4$. It is important to note that they arrived at the correct C^7H^6O minimum formula only because they were using the correct atomic weight values of carbon, hydrogen and oxygen. Had they, like many chemists, been using the incorrect atomic weights of 6 for carbon and 8 for oxygen, they would have arrived not at the correct empirical formula $C^7H^6O^2$ but at $C^7H^3O^2$. This could not be corrected by more accurate work.

There was an intellectual quagmire. It is no surprise that the majority of practical everyday chemists preferred to skirt round it on hard ground and get on with their experiments, using whatever table of atomic weights was closest to hand. What is more surprising is the lack of consensus among the best chemists of the time that this was *an absolutely fundamental problem that had to be sorted out first*. This is especially surprising given that someone *had already sorted it out*.

By about 1835, the Parisian chemists, led by Dumas, had more or less abandoned the atomic theory. The evidence for and against was so confused that it had become an article of faith whether you believed it or not. The true church of Dalton's followers still believed in atoms as an underlying reality, and the ratio of believers to sceptics probably stayed much the same as before. But because it had failed properly to rationalize day-to-day chemistry, the atomic theory was put on the back burner. No atoms, no molecules.

EIGHT

An Infinitely Wise Arrangement

God will provide – if only he would provide until he provides.
                                                    Yiddish proverb

Throughout the eighteenth and nineteenth centuries, the development of new ideas to explain chemistry, as well as the related sciences such as physiology, was complicated by the concept of vital force. This was the belief that the functioning of living organisms could not be explained by purely physical laws, and that there must be some special influence at work.

In the Middle Ages, virtually everyone held to a 'strong' version of the vital force idea. If they thought about it at all, people would have said that the universe was operated by God and his angels, and that a plant would clearly be unable to grow, or a horse run, without His continuous say-so. As far as the alchemists were concerned, their experiments were probing the divine creation, and could get them into trouble. The transmutation of metals was intimately linked in the minds of most of them with the resurrection of Christ and other Christian dogma.

With the arrival of the Enlightenment, things got more complicated. Only a very small minority of scientists in the eighteenth century held to the view that science could explain everything. Most of them felt that their business was to reconcile science with the Christian message, but this was not straightforward. When it came down to the details, there was a fascinating, and what must have been to many religious people an intensely disturbing, interplay of these different world views. Trying to shine the spotlight of reason onto vital force was not an easy undertaking. It was taking a toolbox to the divine mechanism. As time went by, there were all shades of belief. Some scientists remained orthodoxly religious, while at the other extreme there were those who thought that vital force was a hitherto unknown kind of physical phenomenon, in theory compatible with the laws of science but so far inexplicable. Except in France, the majority of scientists – like Liebig, for example – were somewhere in between.

What virtually everyone did at first agree on, though, was that there had to be *something*. Surely, the functioning of organisms could not be explained by the known physical laws; that much was obvious. The really fascinating part of the debate comes with attempts to get to grips with all that territory we now call organic chemistry and biochemistry.

The first person to postulate vital force as a physical phenomenon was Galvani, with his well-known experiments on nerves and muscles. He thought that the twitching of the frog's legs were caused by 'animal electricity' or a 'vital spirit' coming from the brain. Within a few years the physicist Volta had overturned this idea. Galvani's 'animal electricity' was electricity, pure and simple, and there was no room for a vital force to explain the way that the nerves operate. They operate by transmission of electricity, although an explanation of how they generate that electricity, involving the pumping of potassium ions across biological membranes, was still a long way away.

Despite this proof that the role of vital force was not played by electricity, the belief that it existed in some shape or form did not go away. Most people thought that only a vital force could explain not only how the human body worked, but also the plants and their mysterious natural products. The closer a profession came to the inexplicable functions of the human body, the more strongly its practitioners believed in vital force; amongst physicians, faith in it was almost universal.

The spirit of the eighteenth century was one of open enquiry, even into matters as personal to God as the way in which He contrived to manage the universe. But trying to make pronouncements on these matters was dangerous for the believer, for he was always in danger of making assertions that could be disproved by new discoveries in science, thus striking at the roots of belief. Much better to be like the fourth Professor of Chemistry at Cambridge, Richard Watson (1737–1816), who held ecclesiastical posts as well as the chair in chemistry. The worldly Watson was elected to his professorship in 1764, despite his own admission that 'I knew nothing of Chemistry, had never read a syllable on the subject, nor seen a single experiment of it'.[1] Perhaps one is being optimistic in hoping for some profound thought from such a placeman on the subject of the relationship between his professed science and his creator, but he was probably being canny. He found no difficulty in compartmentalizing his beliefs as neatly as he packaged up his career. There is a book of nature and a book of religion, and the two are complementary, he said, before decamping to the Lake District, where he became a country squire and absentee bishop of Llandaff.[2] He clearly, and wisely, refused to get drawn into the argument.

Someone who was perhaps not so wise was Joseph Priestley.[3] The most remarkable thing about Priestley was the contrast between his conservatism in matters of science (his tenacious adherence to phlogiston) and his extreme radicalism in matters of theology and politics. This is precisely the opposite to the set of attributes that would not only have made him a more successful chemist, but might also have given him a quieter life. ('Let's shake some powder out of Priestley's wig,' the mob cried on its way to wrecking his house as a sympathizer of the French Revolution, thus forcing his emigration to the United States.)

Priestley got into extraordinary difficulties. 'His religion was built on the sandy foundation of revealed religion, a foundation that was to be undermined by science.'[4] He thought that matter was interpenetrable, because it was composed of atoms that had no dimensions, only the power of attraction and repulsion; two atoms may exist in the same place. He felt that this property of the interpenetrability of matter neatly dealt with the question of whether man exists as a material being. He does and he does not; he is not solid and inert, but neither is he pure spirit, which makes room for the soul in the material world. As for the vital force, he said this was a natural phenomenon along the lines of electricity or magnetism, which, of course, God must also have created. All in all, Priestley's stance reduced entirely to scientific materialism, except for the belief in God as a first cause, the axioms that 'God exists' and 'God is good', together with the caveat that 'God is unknowable'. Like the alchemists, he had to wrestle with the problem of resurrection; the body is material, and its decomposition is a chemical phenomenon, but 'whatever is decomposed may be recomposed by the Being who first composed it'. This will not be a miracle, but according to some as-yet unknown law of nature. The problem, of course, was that Priestley was giving tremendous hostages to fortune. As soon as someone was able to prove that two atoms could not exist in the same place, would this mean that God did not exist?

Priestley's materialism was much mocked. Bishop Berkeley had told the citizens of the eighteenth century that they were all spirit and no matter, and here was Priestley, supposed to be a man of God, telling them that they were all chemistry and no spirit. His crackpot ideas trying to relate chemistry to theology would be disproved and thus may have caused some otherwise devout people to lose their faith. Dr Johnson thought his views pernicious and his writings 'absurd and impious'.[5] The Welsh poet David Davis wrote a neat epitaph for him:

> Here lies at rest
> In oaken chest,
> Together packed most nicely
> The bones and brains
> Flesh, blood and veins
> And soul of Dr Priestley.

Lavoisier's later findings that organic compounds are made up of the handful of elements that also make up water, chalk, ammonia and other mundane substances also pushed belief to its limits. Had the human body been found to contain totally different elements, or even some rare elements, that might have been more palatable. In fact, Lavoisier's analytical techniques would not have been sophisticated enough to detect the tiny amounts of trace elements, such as manganese or zinc, that the body requires.

It was easier for some people of this period to believe in the literal Christian message than to believe in Lavoisier's chemistry. As James Kier, a chemist of the day, wrote to Erasmus Darwin in 1790: 'You are such an infidel in religion that you cannot believe in transubstantiation, yet you can believe that apples and pears , &tc., sugar, oil, vinegar, are nothing but water and charcoal, and that it is a great improvement in language to call these things by one word – oxyde hydro-carbonneux.'[6] So it was the chemist who did not believe Lavoisier!

By this time the majority of people in England, even if they were not devoutly religious, held to vague theism. Priestley's attempt to square religion with materialism was rejected and it was generally held that religion should be based on faith, not evidence. A more typical noncommittal take on vital force is illustrated by the instructional book for females *Conversations on Chemistry*, written by Mrs Marcet (Jane Haldimand) and published in 1806 as a Socratic dialogue between the knowledgeable Mrs B. and her two pupils, Caroline and Emily. The book was immensely popular, selling 160,000 copies in the United States alone, and is said to have inspired the young Michael Faraday to take up chemistry. The first volume is an adequate account of the current state of Lavoisier's inorganic chemistry. Most of the second volume deals with the products of living organisms and hardly touches on what we would now call chemistry at all; it is a description of milk, sap, bitumen and all the other complex products of the natural world of which current knowledge was less than rudimentary. The author says that 'organised bodies bear the most striking and impressive evidence of design, and are eminently distinguished by that unknown principle called *life*'. When Caroline enquires how life enables these organs to perform their several functions, Mrs B. has to throw up her hands: 'This is a mystery which, I fear, is enveloped in such profound darkness that there is very little hope of our ever being able to unfold it.'

The subject is then dropped until the book's very last paragraph, when the reader is enjoined to remember that both the spontaneous operations of nature and her own powers of thought proceed from the goodness of Providence: 'To GOD alone man owes the admirable faculties which enable him to improve and modify the productions of nature, no less than those productions themselves . . .'. The muddling of God with Providence seems typical of the age, and the last paragraph is just going through the expected motions.

In 1829 the Revd Francis Henry, Earl of Bridgewater, died leaving the sum of £8,000 in the care of the President of the Royal Society. He was instructed to use the money to engage persons to write and publish 1,000 copies of a work *On the Power Wisdom and Goodness of God, as Manifested in the Creation*. After due deliberation, the President came to the conclusion that the earl's wishes would best be fulfilled by a series of eight monographs. One of these, on chemistry, meteorology and digestion, by William Prout ran to four editions between 1834 and 1855.[7]

Prout's book is closely argued, and he does not give the impression of a man using the earl's bequest as a gravy train. His opponents, he says, are those whose 'minds are so obtuse, or so singularly constituted, that they maintain all the appearances of design to be unreal'. Chemistry is full of mystery. The chemist, he says, is in the position of a non-mechanical onlooker who stands by a threshing machine or some other contrivance, and whose view of the interior is obscured by the casing. He has no idea how it works, but he can see that certain materials passing through it undergo remarkable changes – changes devoted to an end and arguing for the conscious design of the machine. The chemist may well, out of curiosity, wish to remove the casing and look inside; but this may not explain to him how it works, and he does not need to dismantle it to convince himself that it was deliberately built. The obscurity of the way in which the machine operates is, in fact, the best argument for design, for, when the Deity operates through the medium of a clearly understood mechanism, 'He appears almost too obviously to limit his powers within the trammels of necessity'. It is precisely because chemistry is so mysterious that His powers appear of a higher order. Prout, unlike Priestley, believed in an immediate deity, that is one who, through the intermediary of delegated agents (Prout does not use the term 'angels'), runs everything on a second-by-second basis, like a kind of superefficient Josef Stalin.

Organic chemistry, being the most mysterious branch of the science, is therefore the one in which the hand of the Great Author of Nature is most clearly seen. The number and diversity of organic substances appear to be endless, and God has here chosen to manifest his attributes of infinity. But the use of only a few elements in the creation of all these different acids, alkaloids and so forth is even more convincing: 'Instead of different principles, the same carbon, nitrogen, the same water enter into every living being. We have often thought that the Deity has displayed a greater stretch of power in accommodating to such an extraordinary variety of changes a material so unpromising and so refractory as charcoal.' The only slight note of uncertainty comes when he considers the more lumpen discipline of inorganic chemistry, where he has to admit that the purpose of the creator in throwing together some of the elements is obscure. The use of such as niobium in the economy of nature is at present beyond our knowledge, he says. It has to be admitted that, even to the present day, a number of the metallic elements have no appreciable practical role except that of keeping inorganic chemists amused.

Berzelius too believed in a vital force that made organic compounds special. This was linked with the belief, inherited from Lavoisier, that their synthesis was an impossible aspiration. Nature made the organic compounds from inorganic materials. The chemist could not; he could only degrade them. Nature makes, man breaks, as Lavoisier had said. As well as being incorrect in fact, this idea had the consequence that, until about the middle of the century, most chemists thought that useful information about organic compounds could be gained only by their degradation, and that attempts at synthesis were a waste of time. Because of this, Berzelius and others tended to treat the natural products in a descriptive kind of way, like plants, insects, and the other products of Nature's bounty. They became disconnected from the physical sciences, which at the time only embraced mathematics, what we would now call physics, and inorganic chemistry.[8] This approach, following Fourcroy, had its grounding in the vital-force idea.

As Hofmann described it half a century later, vital force became a kind of bugbear on scientific research. There was a 'reluctance felt by the inquirers of that period to divest any process, however simple, when accomplished within the living organism, of the operation of vital force'. It was thought to modify and overrule in a mysterious way all biochemical actions.[9]

Another related scientific debate that was taking place in mid-century concerned fermentation. Liebig was at the head of those who thought that organic compounds were inherently unstable, and that contact with yeast, for example, could cause their breakdown, a process known as 'eremacausis'. It was a catalytic process analogous to the way that a platinum wire could cause an infinite amount of hydrogen and oxygen to combine without itself being changed. Liebig remained to the end a ferocious opponent of those who believed that the yeast was a living organism actually metabolizing sugar into alcohol. In the last paper he published before his death in 1873, he compared them to the man who thought that the Rhine was driven along by the rotation of the string of water mills across the river at Mayence.[10]

Some thought that this catalysis arose through the exercise of a 'catalytic force', which would imply a close relationship with vitalism. Liebig and others (including Berzelius) preferred to see it as a purely chemical process. Diseases, too, were examples of eremacausis, the uncontrolled breakdown of the tissues catalysed by unfavourable chemical circumstances. Neither Liebig nor the doctors yet knew the importance of micro-organisms, but, whereas most chemists rejected the vital force and saw eremacausis as a purely chemical phenomenon, there was ample scope for the interpolation by others of vital force into these mysterious processes.

Ironically, it was Prout who first correctly analysed the substance, urea, that was to play a major role in the demolition of the vital-force theory. Urea is a major constituent of urine, from which it was first isolated by Vauquelin. It is clearly a natural substance and, therefore, according to the vitalists, must be a reservoir of vital force.

In 1827 or 1828 Wöhler carried out the experiment for which he is now most famous, again based on his cyanates.[11]

At this point we need to skip ahead again in time to explore the structure of these cyanates in a way that was not yet accessible to Wöhler and his contemporaries. We have seen how a reaction between a halogen and an alkali metal produces a salt, such as sodium chloride, in which an electron from the alkali metal sodium is transferred to the halogen chlorine, converting the atoms into positive and negatively charged ions respectively. Cyanates are salts in which a small group of atoms (one nitrogen, one oxygen and one *carbon*) mimics the chloride ion.

Lavoisier had first used the term 'radical' to describe a group of atoms that behaves as one and passes unchanged through a reaction or series of reactions. The cyanate group, consisting of three atoms, is a small radical, hardly bigger than chloride, that behaves as a negatively charged body mimicking in its chemical behaviour many of its properties. The term 'pseudohalogen' was coined later to describe cyanate and some other similar groups.

Just as cyanate mimics the halogens, so certain small, positively charged radicals can mimic the alkali metals. Chief among these is the ammonium ion, four hydrogens surrounding a single nitrogen, which is about the same size as the potassium ion and carries the same single positive charge. Ammonium salts (for example, ammonium chloride or sal volatile) resemble in many of their properties the salts of the alkali metals.

All these structural insights were in the future when Wöhler carried out his 1828 experiments on cyanates. By reacting silver cyanate with ammonium chloride he was attempting to prepare the ammonium salt of his cyanic acid. But, when he filtered off the insoluble silver chloride and evaporated the clear solution left behind, he obtained 'colourless, clear crystals in the form of slender four-sided dull-pointed prisms', rather unreactive, but seeming to resemble what he knew of urea.

'This similarity . . . induced me to carry out comparative experiments with completely pure urea isolated from urine, from which it was plainly apparent that [urea and] this crystalline substance, or cyanate of ammonia, if one can so call it, are completely identical compounds.'

Wöhler wrote to Berzelius: 'I cannot, so to speak, hold my chemical water, and must tell you that I can make urea without the necessity of kidneys, or even of an animal, whether man or dog . . . ammonium cyanate is urea.'

Wöhler was incorrect to state that the two substances are the same. He goes on to say more correctly that 'two different bodies can contain the same proportions of the same elements, and that only the different kind of combination produces the difference in properties'. In other words, ammonum cyanate and urea are not the same, but isomers, like the fulminates and cyanates, differing in the arrangements of their atoms. In this experiment, ammonium cyanate is first produced in solution, but isomerizes to urea before the solution can be evaporated. This is how this can be written in modern notation.

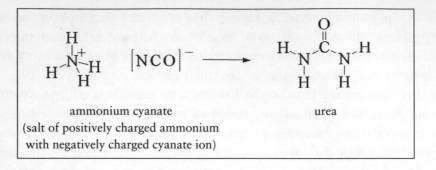

ammonium cyanate
(salt of positively charged ammonium
with negatively charged cyanate ion)

urea

Two years later, in 1830, Wöhler and Liebig successfully prepared solid ammonium cyanate by working in anhydrous conditions, and obtained it as a white powder that was stable in a sealed vessel, but isomerized to urea on exposure to the atmosphere for a couple of days.

Seen from a modern viewpoint, the chemistry involved in Wöhler's experiment is very simple and of no great significance. But his was one of those iconic experiments, like Newton's supposed apple, that has entered mythology. He had made a substance, urea, hitherto known only as a product of animals, from inorganic substances by simple means and without any intervention of vital force.

According to the modern definition, organic chemistry is the chemistry of carbon, but very simple carbon compounds like carbon dioxide and calcium carbonate (chalk) are usually considered part of inorganic chemistry. Ammonium cyanate already contains carbon, so, although resembling sodium chloride in some of its properties, it is by some definitions already organic. Urea, too, is a particularly simple organic compound, hardly bearing comparison for complexity with carbohydrates or proteins. In modern terms, Wöhler was patrolling the boundary between organic and inorganic chemistry through a region in which the frontiers were ill defined. Wöhler actually claimed less for his synthesis than he might have done, for he stated that 'it is remarkable that for the preparation of cyanic acid (and also ammonia) an organic substance is always originally necessary'. He had overlooked the fact that Priestley had synthesized ammonia from nitric acid in 1781 and that nitric acid had been synthesized from its elements by Henry Cavendish in 1785. He also failed to notice that, on the other hand, potassium cyanide, from which he had made his cyanates, had also been made from its elements.

In fact, his was not strictly speaking even the first synthesis of an 'organic' compound. In 1784 the brilliant Scheele had made hydrogen cyanide by heating to redness a mixture of potassium carbonate and charcoal, then adding ammonium chloride. At the time, it was not known that hydrogen cyanide, or prussic acid, was a natural product, but it is present in apricot kernels and other fruit stones, both free and combined with sugars. A dozen or so of the little 'nuts' found inside apricot stones is a fatal dose for a small child. In 1784 the boundary between 'organic' and 'inorganic' was still unclear; Lavoisier had not yet defined organic compounds as being the compounds of carbon. The ammonium chloride, or sal volatile, that Scheele used could be defined as 'organic' because at that time it was extracted from the salts that crystallized out around animal manure.

Even in 1828, the science of organic chemistry was only just emerging and 'organic' meant compounds such as urea from living organisms. Viewed from this perspective, there are no doubts about the significance of Wöhler's experiment. A natural product had formally been made from its constituent elements. Liebig and his successors regarded the experiment as marking the first beginning of a truly scientific organic chemistry. The way was cleared for the chemists to beat their path into the heartland of the natural products.

What, then, of the vital force? Wöhler's experiment did not banish the idea overnight, despite what some subsequent writers have said. While there is no doubt about what he actually did in 1828, opinions continued, and still continue, to be divided about its exact significance. Berzelius for one never wholly abandoned the idea of vital force until his death in 1848. 'The living individual, which dies and gives its ingredients back to the inorganic world, is not recreated. The nature of living bodies is thus not grounded in their inorganic elements but in something else . . . what it is, where it comes from and where it goes, we cannot comprehend.'[12]

In 1827 Berzelius had explained the phenomenon of isomerism between cyanic and fulminic acids using exactly this mysterious phenomenon. Two compounds of the same composition can have different properties only as a result of the different action of the vital force. He had to take a half-step back from his earlier assertion that the synthesis of organic compounds was impossible, and accommodated Wöhler's urea synthesis by stating that it was possible only for certain substances, which he said stood on the extreme limit between organic and inorganic. This represented at least some advance on the tenets of *Naturphilosophie*, which held it improper to attempt to apply the laws of the inorganic universe to living systems. Berzelius's views on vital force were intimately related to his difficulties in trying to force the round peg of organic compounds into the square hole of his electrical theory of bonding.

Liebig's views are of interest, since from about 1839 on he began important research on physiology and the functioning of the animal organism, and became a highly influential figure outside the scientific establishment as well as within it. At first he believed in a vital force and considered it to be a natural physical phenomenon, not the same as electricity, or magnetism, or chemical reaction, but a 'peculiar force' whose effects are produced by no other. Natural science has fixed limits that cannot be passed.[13] Brought up a Lutheran but with a Catholic wife, he seems to have been the kind of middle-of-the-road theist who would have been welcomed into the bosom of the Anglican Church, and at first saw no major discrepancy between these beliefs and his chemical work. With advancing years, he persisted with these relatively unsophisticated beliefs, which by then were becoming unfashionable among scientists. 'Our concept of God changes, purifies and becomes enlarged with our idea of "force",' he wrote in 1867.[14] The seventh edition of his *Agricultural Chemistry* (1862) caused embarrassment not only on account of its libellous tone, but because of its references to a Creator, which his friend Mohr said gave hostage to the materialists. This is despite the fact that certain of Liebig's physiological experiments knocked further holes in the vital-force theory. Following in the steps of Lavoisier, he did some experiments on food consumption and heat production in animals. He established conclusively that all the energy of the organism is produced by combustion of food in the tissues, and there is no need for other mysterious sources of motivation such as electricity, or 'nervous energy', or vital force.

Wöhler's urea synthesis was only the first nail in the coffin of vital force, but, as far as organic chemistry was concerned, the concept rapidly faded. Wöhler and Liebig isolated and characterized several further constituents of urine, and believed that science would one day be able to synthesize all the products of the living organism. They were effectively founding the science of biochemistry, albeit still at a primitive stage. In 1845 Kolbe synthesized acetic acid from carbon tetrachloride CCl_4 (which he thought C^2Cl^2). Marcellin Berthelot declared it as one of his main scientific aims to abolish the concept of vital force, and in his numerous research papers reported the syntheses of many more organic compounds, finally putting to flight the notion inherited from Lavoisier and Berzelius that only Nature could make them. Albert Ladenburg eventually reported the first rational synthesis of an alkaloid, coniine, although not until 1886.

Other scientists, though, were not convinced. As greater knowledge was gained, especially with the microscope, of the astonishing complexity of the living world, what actually happened was that the theists moved their tents down the road towards the biological. The earlier vitalists, they now said, had made the mistake of claiming that all organic chemistry was due to vital force. They had 'claimed for that agency [vitalism] much that is clearly attributable to the operation of chemical and physical forces'. This could not be maintained. But 'living bodies present a large class of phenomena which are altogether peculiar to them, and which can only be attributed to agencies of which the inorganic world is altogether independent'.[15] Simple compounds might be capable of synthesis using heat, light, electricity and so forth, but Carpenter, talking about the growth of plant seeds, subtly shifts the definition of an organic compound: 'The germ, under the influence of light, decomposes carbonic acid [carbon dioxide], and unites its carbon with the elements of water . . . thus forming organic compounds, *such as no operation of ordinary chemistry has yet been able to imitate.*'[16]

There is no mention of Wöhler. The sentence is ambiguous, and can be read in either of two ways. Either an 'organic compound' is now defined as anything that the chemist cannot synthesize, or else it is defined as anything that the chemist cannot synthesize *in the same way that the plant does*. The 'growth, multiplication and transformation' of living organisms indicates, 'in the most distinct and unmistakeable manner, the controlling and sustaining action of an intelligent mind, acting in accordance with a determinate plan'. According to this author, the idea that the 'organizing force' that is responsible for the development of the whole organism lies dormant in something as minute as a single cell 'seems almost too absurd ever to have gained believers'. Carpenter, a physiologist and anatomist, believed, in harmony with Locke, that 'all force which does not emanate from the will of created sentient beings, directly and immediately proceeds from the Will of the Omnipotent and Ominipresent Creator'.[17]

Fifteen years later, his original speculations having on his own admission generated little debate, Carpenter published an updated version.[18] This time there is no mention whatsoever of the Creator, although he still believes in vitalism. He seems to have abandoned the ground of chemical synthesis to the chemists, but is still unable to accept the possibility that an embryo can divide and grow without the continuous infusion of a mysterious external force.

It might appear that the vitalists were now in full retreat, but they would make a reappearance later, in a most unexpected manner, and through the influence of further studies of the very science, organic chemistry, that had seemed to put them to flight.

Liebig did not live to see this. In 1870, not long before his death and following a grave illness, he ordered his coffin and wrote to his sister: 'my familiarity with Nature and her laws has convinced me that we should not worry ourselves about death and our future life since everything is so infinitely wisely arranged that anxiety about what happens after death has no place in the soul of the scientist. Everything has been cared for, and whatever becomes of us is surely for the best.'[19]

NINE

New Bodies Floating in the Air

CAROLINE: *To-day, Mrs. B——, I believe we are to learn the
nature and properties of CARBONE. This substance is quite new to
me; I never heard it mentioned before.*
Mrs Marcet, *Conversations on Chemistry, in which the Elements
of that Science are Familiarly Explained and Illustrated by
Experiments*, 2nd edn (London, 1807)

By 1816, twenty-six miles of piping had been laid in London in
order to carry illuminating gas to factories and street lamps,
although few homes had gas lighting until the second half of the
century. When the gas mains did eventually spread out to domestic
premises, the first areas to be connected would have been what
were then the inner suburbs, such as Newington in Surrey, today an
inner-city area, where Michael Faraday was born on 22 September
1791.

Faraday's father was a blacksmith, originally from Yorkshire. The
relative poverty of the family meant that by the age of thirteen his
son was working at a bookbinders, where he fell into the habit of
reading the books sent in. One of these was Mrs Marcet's
Conversations on Chemistry. Soon he was carrying out chemical
experiments and attending Davy's lectures, then corresponding with
the famous chemist. By the time he was twenty-one, he was Davy's
laboratory assistant at the Royal Institution.[1]

Faraday was devoutly religious, absolutely sincere and truthful,
but not pious; he had a sense of humour and was good company,
with a somewhat excitable nature, although he invariably worked
alone. All in all, no one seems to have had a bad word to say about
him, although, as far as chemistry was concerned, he had one major
failing; he did not believe in the atomic theory. Like Davy, he

thought that Dalton's ideas were a very clumsy hypothesis. Going even further, he rejected almost everything that would make chemistry real: nomenclature, atomic weights, equivalent weights, what have you. Given this, it is perhaps not surprising that his chemical researches, though interesting enough, do not bear comparison with his later work on electricity and related subjects. With one exception.

Early gas supplies were not even made from coal. At the gasworks, whale oil or cod oil was poured into a red-hot furnace, producing a hideous mixture of products that was put into iron vessels under pressure. These iron cylinders were taken by horse and cart to the premises of the subscriber, where the cylinder was connected up to his gas pipes.

A problem that the gas suppliers had, not to mention the appalling environment of flame, cancer-causing hydrocarbons and horse manure in which they had to work, was that the gas contained a high proportion of liquid products that would condense out at the works or in the pipes. The volume of this so-called gas oil was considerable; about one gallon for every 1,000 cubic feet of gas.

In 1825 the owners of a gasworks in London sent Michael Faraday a sample of this liquid. The initial contact probably took place through his brother, who was working for a gas company, although Faraday was also involved as an expert witness in an insurance case involving a fire caused by burning oil.[2] On 9 May, his notebooks record: 'Went to Day to Portable Gas works, saw Mr Gordon and the Arrangements, Mr Turnbull, engineer. Mr Hasledon superintends the filling, etc.'[3]

Faraday experimented with the liquid gas oil for ten days or so. It was clearly a mixture of substances, because, when carefully distilled, it boiled over a range of temperature; here we see Faraday applying careful physical methods to study the purity of his substances, methods developed by the Parisian chemists. Eventually, he obtained, by a combination of distillations and freezing, a liquid boiling at 85.5°C that appeared to be homogeneous.[4] He found it to be a hydrocarbon – that is, a substance containing carbon and hydrogen only – and, using the current atomic weight 6 of carbon (that is, the carbon atom was thought to be six times heavier than that of hydrogen), he assigned the formula C^2H. Faraday's experimental figures were accurate, but the formula is wrong in two respects. First, the atomic weight of carbon is 12 not 6, so the formula of the liquid is CH, or a multiple thereof, not C^2H; and, secondly, the 'multiple thereof' is six: the correct molecular formula of the liquid is C_6H_6. This formula is supported by accurate measurements of the vapour density of the substance made by Faraday himself; he found it to be 'nearly 40' (forty times the value for hydrogen); the correct value for C_6H_6 is 39; but the relationship between vapour density and formula was at that time obscure.[5]

Faraday named his liquid 'bicarburet of hydrogen'. It was to remain a chemical curiosity for the time being. There was no indication of the central role that benzene, as it was later named, was to play in chemistry, or of the oceans of human ingenuity that were to go into explaining its true nature during the course of well over a century.

Neither was it known at this time that benzene is the parent substance of the vast series of substances known as the aromatic compounds. Like the word 'organic', the word 'aromatic' underwent a change of meaning in the nineteenth century, as an underlying scientific classification emerged to unite substances formerly related by a much more subjective property – in this case, their pleasant odour. Benzene itself, as it happens, is not a particularly pleasant-smelling liquid, and is actively toxic. According to an anecdote, Hofmann, the famous pupil of Liebig who became an important chemist in the later part of the century, invariably described in his lectures how a certain lady had told him that the smell of benzene reminded her of washed gloves. On one occasion, a student in the back row who had heard him lecture before broke in with the phrase before Hofmann could utter it. 'Like washed gloves', the heckler called out. 'Oh,' replied Hofmann. 'You knew the lady also, then?'[6]

Many of the derivatives of benzene had been known since ancient times as present in pleasant-smelling oils and spices: vanilla, cinnamon and many others. The name 'benzene' was derived by Liebig from that of benzoic acid, which in turn was obtained from gum benzoin, a product of the East Indies. The name for the gum appears to be derived from an Arabic name Lubān Jāwi, or incense of Java, later corrupted by Portuguese merchants into benjawi, thence into benzoin, benjamin and so on. It was used especially in church incenses.[7] Benzoic acid had been obtained from the gum as early as 1557, and was studied by Scheele.

The discovery of benzene, or 'bicarburet of hydrogen', more or less marked the culmination of Faraday's researches in pure chemistry. From about this date onward, he concentrated on his ground-breaking work in electricity and electrochemistry, for which he is most famous. The sample of benzene that he had placed in a sealed tube sat on the shelf as an unexplained curiosity. Many years later, it was produced with pride at the famous banquet in Berlin, and occasionally ever since, when the occasion seems to merit it.

Seven years after this excursion of Faraday's, in 1832, Liebig and Wöhler began to work together again, this time on the aromatic compounds, researches that were to result in their second great joint paper.[8]

They began by isolating from oil of bitter almonds an oily substance that they called hydrobenzoyl, containing carbon, hydrogen and oxygen only. Analysis gave the following figures: carbon, 79.4 per cent, hydrogen 5.8 per cent and oxygen (by difference) 14.8 per cent. On the basis of this analysis, they assigned the formula $C^{14}H^{12}O^2$ (twice the correct figure of C_7H_6O). Earlier workers had shown that, when bitter almond oil oxidized in air, it formed the crystalline substance benzoic acid, also obtainable from gum benzoin and other plant sources. Liebig and Wöhler confirmed that pure hydrobenzoyl (today called benzaldehyde) was oxidized by air to benzoic acid, which they also analysed, finding it to be $C^{14}H^{12}O^4$ (correct figure $C_7H_6O_2$) and correcting an earlier analysis by Berzelius.

They then carried out a series of further transformations on hydrobenzoyl. Treating it with chlorine gave benzoyl chloride, $C^{14}H^{10}O^2Cl^2$, a reactive substance that could be further converted by potassium iodide into benzoyl iodide $C^{14}H^{10}O^2I^2$ and by ammonia into benzamide, $C^{14}H^{14}O^2N^2$. Benzoyl chloride and benzoyl iodide both formed benzoic acid on treatment with water.

This was the first series of chemical transformations carried out within a group of such closely related organic substances. Their paper is a model of presentation and has a claim to be the first modern chemical research paper. The facts are clearly and succinctly set out, and the conditions for each reaction are clearly given, together with the properties of the new substances. For once, the experimental work is not mixed up with speculations, diversions, accusations and polemics.

The two realized that, in this series of transformations, a large part of the molecular formula, which they formulated as $C^{14}H^{10}O^2$, remained unchanged. This grouping, which they called 'benzoyl', they considered to be a compound radical. The concept of a radical, or collection of elements mimicking the behaviour of a single element, had first been advanced by Lavoisier, but in a much simpler form based on inorganic examples. Their findings were praised by Berzelius, to whom they sent a copy before publication. 'The results . . . are the most important that have yet been obtained in vegetable chemistry . . . one can see a new era.' He approved their proposed name 'benzoyl' and at first enthusiastically applied the radical concept to other known reactions. His paper, printed immediately after theirs, does, however, finish with the caveat: 'These formulae should only be promulgated when the ideas that you have put in print have become somewhat more established; otherwise, only a Babylonian tangle will result.'

These reactions became even more significant when the Berlin chemist Mitscherlich carried out an important further transformation that Liebig and Wöhler had missed.[9] By heating benzoic acid with lime in 1834 he obtained nothing other than Faraday's 'bicarburet of hydrogen'.

The significance of his results was, however, not immediately clear, because they set in motion another vast argument. Liebig wrote an 'Addendum from the Editors' on Mitscherlich's results, which at nine pages exceeded the original paper in length and which set out to inform the world what Mitscherlich had really done. Liebig did not get on with the hypersensitive Mitscherlich, and his polemic was 'pedantic at the very least', full of phrases like 'Surely this sentence means that . . .'.[10]

The question was, what was the true 'radical' of the aromatic compounds? Liebig was clearly vastly annoyed that Mitscherlich and not he had discovered the conversion of benzoic acid to benzene, and continued to regard his benzoyl as the true radical. Benzene was merely a kind of 'rubble-fragment' that Mitscherlich had stumbled upon. If benzoyl is the true radical, then benzene is 'a kind of extreme radical-form [*ultraliberalen Radikal-gewand*], if one can coin the phrase'. It is just possible that Liebig may have been making a joke here, perhaps about politics or even his own past; 'ultraliberal' and 'radical' have obvious political meanings, and *Gewand* can mean a cloak as well as a shape. But the overall tone is bad-tempered and intolerant. There had been simmering discontent between them ever since Mitscherlich had criticized Liebig's teacher, Gay-Lussac, years before. Now Liebig wrote to Wöhler that he felt relieved that it was now open warfare; he wrote to Berzelius that 'Mitscherlich gorges himself only on foreign fat; he is a leech with respect to everything in his vicinity, and once he has sucked to his heart's content he discards the victim like a squeezed lemon. . . . This lurking, malicious being, this half-cursed friendship that betrays me like Judas while at the same time expressing its affection for me, is repugnant to me to death . . .'.[11] There was no one correct answer, and the vitriolic exchanges were totally unjustified by the facts; benzoyl and benzene can both be regarded as radicals, and it is a matter of definition; but, if anyone was correct in this dispute, it was Mitscherlich. Benzene, C_6H_6, is the true irreducible nucleus or radical of the aromatic compounds; benzoyl is the larger fragment present in many compounds, in which the benzene or phenyl radical C_6H_5 is attached to a carbonyl group: $(C_6H_5)–CO–$.

Berzelius was fond of Mitscherlich, who had been his student even before Wöhler, and who worshipped him. It was difficult for Berzelius to take sides in this dispute, and it may have been difficult for him to unscramble ideas about the chemistry from the clash of personalities. Whatever the underlying motivations, within a short time he had got cold feet on the subject of radicals. He began to feel that the idea of a radical containing three elements was against the concept as initially enunciated by Lavoisier, and it was difficult to reconcile with his own electrochemical theory of constitution. In particular, how could benzoyl chloride and benzoyl iodide, containing the electronegative halogen elements, be intimately related to hydrobenzoyl (benzaldehyde), containing the electropositive hydrogen? He also criticized Mitscherlich's ideas on benzoic acid, notably his suggestion that not only might it be a compound of benzene with carbon dioxide, but that all organic acids might be compounds of carbon dioxide with a hydrocarbon. This is close to the truth, and would have been a most productive line of thought if properly exploited. But once again Berzelius's great prestige had an inhibitory effect. The result was that the hypersensitive Mitscherlich, wounded by the criticism, wrote to Berzelius saying that he wished he had never heard of either benzene or benzoic acid.[12]

Ignorance about what benzene was did not prevent it becoming an important source of new substances. By mid-century, Hofmann's students in London in particular were obtaining large amounts of benzene from coal or gas oil, and converting it by two practicable steps into aniline, another simple and versatile benzene derivative that became the building block for many dyes and other intermediates. Previously, aniline had been known only as a breakdown product of the extremely costly indigo, but now here were young men in frock coats (in Oxford Street no less!) nitrating benzene with a mixture of nitric and sulphuric acids, then treating the resulting nitrobenzene with iron powder to produce it in gallon quantities. Painstaking research by Hofmann and others built up a large network of interrelated substances, all containing the benzene radical or nucleus. Hofmann was undeterred by the fundamental uncertainty about the substances they were working with. 'Gentlemen, new bodies are *floating* in the air,' he said one day, when yet another inexplicable new reaction was discovered.[13]

But Hofmann was exceptional at this time. The older scientists despaired; in 1852, Faraday, who was no longer actively engaged in chemical research, wrote: 'I am too stupid to understand the organic chemistry of the present day and I will acknowledge that through my bad memory organic chemistry is to me a sealed book.' Another contemporary thought that chemistry was being frittered away by the hairsplitting of the organic chemists: 'We have new compounds discovered which scarcely differ from the known ones and when discovered are valueless . . . very little acting to the progress of true science.'[14] By about 1840, Liebig had largely given up organic chemistry for agricultural chemistry and nutrition, and at about the same time even Wöhler more or less abandoned the subject for inorganic chemistry.

As more and more data about the properties of different organic compounds accumulated, it became increasingly clear that benzene was wholly exceptional. All other known substances with such a low ratio of hydrogen to carbon were unstable. Benzene had the same ratio, 1:1, of the two elements as the highly reactive and explosive gas acetylene, but there was no sign of the instability that such a formula seemed to demand. It was physically stable, burning in air with a smoky flame, but was otherwise difficult to get into reaction. The benzene nucleus, as Liebig and Wöhler had demonstrated, could pass unchanged through whole series of substitution reactions that could lead back again, given the right manipulations, to benzene itself. It was an enigma. The eventual solution of the nature of benzene would be one of the great attainments of the human mind.

TEN

Adieu, Monsieur!

If ever a man elevated himself by the force of genius and perseverance, such pre-eminently was Laurent, to whom everything was denied, and who made his chef d'œuvre *out of nothing.*

Correspondence of M. Jerome Nicklès (April 1853), cited by Novitski (1992)

The intellectual life of Paris is brilliant only superficially, a kind of champagne, but one that leaves a bitter taste.

Albert Jacqard, quoted by Theodore Zeldin, *The French* (1983)

For the first seventy years of the nineteenth century, France was intermittently a violent country. Henri Regnault, who studied with Liebig, was orphaned at the age of two when his father, an engineer officer in Napoleon's army, was killed in the invasion of Russia. Regnault's son was in turn killed in the Franco-Prussian war of 1870, when Regnault's laboratory at Sèvres was deliberately wrecked by Prussian troops and his papers all destroyed.

In between these two wars, the country remained turbulent. Following the three-day revolution in 1830, there were violent insurrections in Lyon and Paris. On 15 April 1834 the inhabitants of the Rue Transnonain blockaded the street against government troops and were massacred. By 1848 the barricades had been up in Paris eight times during the previous two decades; in that year came the February revolution. This brought Louis Napoleon to power, first as President of the Second Republic, then in 1852 as emperor with more absolute power than any other monarch in Europe.[1] The chemist Cahours wrote to his friend Gerhardt on 22 February 1848: 'Paris is on fire; a hundred thousand insurgents on the streets, troops stationed everywhere, one no longer knows what will happen.'[2]

Against this background, it is hardly surprising that personal relationships could be violent too. The 1840s witnessed a turbulent interaction between the theories, and the personalities, of three French chemists. Jean-Baptiste Dumas, a member of the French establishment, was the senior of the trio; Auguste Laurent and Charles Gerhardt the two rebels. Their story has to be seen against the backdrop of the fortunes of their country, and gives some fascinating insights as to how France began to fall behind Germany in the race to the new science. The full story of the remarkable partnership between the last two has not been written. Important questions remain unresolved; which of the two was the greater, and did Gerhardt betray his friend after death?

The country was kept under control only by iron centralization. Whoever controlled the capital controlled France, and a Parisian elite controlled scientific life, as it controlled every other aspect of national life. Those like Laurent and Gerhardt unfortunate enough to hold posts in the distant provinces felt, if they were any good, that they were wasting their time. Coteries, personalities, politics could inflict grave wounds on the *amour-propre* of the participants and warp their judgement; to their supporters, the two were outsiders and suffered from the indignity of exclusion from the Parisian in-group. Gerhardt's biographers record indignantly that, when the President of the London Chemical Society, Thomas Graham, visited Paris in 1851 along with Liebig's acolyte A.W. Hofmann, Thenard gave a dinner at his house in Fontenay-aux-Roses in their honour. Many bottles of wine were dispatched, including a *clos de Montrachet* that Hofmann recalled with pleasure. All the chemists were there, Thenard recorded fondly, with three exceptions: Chevreuil, who appears to have been a valetudinarian and sent his excuses, and 'DES DEUX', not even mentioned by name.[3]

This metropolitan backbiting was in marked contrast to Germany, where there was friendly, or sometimes less-than-friendly, rivalry between the numerous scattered laboratories that encouraged the development of new ways of doing things. In France, there tended to be one set of received scientific opinions at any one time, honed by discussion in Paris. This could have its disadvantages. When, for example, around 1835, disillusionment spread about the atomic theory, the Paris chemists, led by Dumas, largely dropped it as a group.

Dumas was originally intended for a career in the navy; the defeats of 1815 made his family change their minds and he became apprenticed to an apothecary.[4] The following year, Dumas set out to walk from Alais where he had been born to Geneva, where he studied biology as well as chemistry. There, with Berzelius, he became one of the first people seriously to question the usefulness of the gross analyses of biological materials that had been done until then. What was the point of carrying out numerous analyses of whole milk, for example, without taking into account the composition of the fatty globules that it contained, he asked. This work attracted attention.

In Geneva he came, like Liebig, under the spell of the charismatic Humboldt, who turned up wearing the dress of a dandy of a previous era, complete with shiny metal buttons. Will you be my cicerone?, Humboldt asked him, an offer that Dumas found irresistible. It was Humboldt's descriptions of 'la vie Parisienne' that resulted in Dumas finding his way to that city.[4]

Twenty years later, he had become a grandee. Gerhardt's biographers describe him receiving the penniless young chemist on his arrival in Paris with the magnificent pride of a grand seigneur,[5] proud above all of the figure he cut, his grand manners, and proud of giving advice that was vague and banal, with no hint of practical help. Louis Pasteur described him giving his packed lectures like a svelte advocate, dressed in black suit with white shirt and cravat; only knocked off his stride when his assistant one day prepared a demonstration incorrectly. He continued to rise seamlessly through the French establishment, again, according to Pasteur, 'passing twenty years of his life receiving honours, neither soliciting them nor showing any sign of surprise'.[6]

A Roman Catholic, Dumas was a talented experimenter with whom Liebig and Wöhler had a kind of knockabout relationship. Wöhler called him a 'Jesuit' and a 'windbag' but recognized his talent. Liebig wrote: 'It always annoys me that this fellow, in spite of his unclean, impossible and bad way of working, yet with the Devil's help fetches masterpieces out of his sleeve.' Adverse criticism of French scientists for working sloppily and publishing too quickly was generally widespread around this time.[7] Dumas probably realized that the old way of doing scientific research was causing France to slip behind. He was the first chemist in France to try to follow Liebig's example in giving practical laboratory instruction to his students, at his own expense, but in 1848 his income was so affected by the Revolution that he had to abandon the classes.

Auguste Laurent (christened Augustin) was born in a village in the Haute-Marne in 1807, the son of a wine merchant who insisted that he learn bookkeeping.[8] But the young Auguste, like Berzelius, preferred to wander the countryside studying rocks and plants. On the advice of his teachers, his father bowed to the inevitable and the youth went to study at the prestigious École des Mines in Paris. In 1831 he became assistant to Dumas, shortly to become Thenard's successor as professor at the École Polytechnique. Not long afterwards, Laurent was appointed analytical chemist at the Sèvres porcelain factory, under the direction of Dumas's father-in-law, Brongniart, but he found the work tedious; he had been bitten by the bug of organic chemistry, and resigned to found his own private school of chemistry in a Paris garret. He had no interest whatsoever in money, except when its absence made it impossible for him to work properly, a constant theme of his later letters. When he found that teaching the students was taking up too much of his time, he told them to go away, and he lived in penury rather than cut down on his beloved research. For a period he worked in a perfumery shop, where the owner let him do his own experiments. Laurent was apparently astonished later to be paid 10,000 francs for minding the shop. A long and rambling 1837 paper by him entitled 'Diverse Researches on Organic Chemistry' starts pathetically by stating: 'I had hoped that the labours so far published by me would have opened the door of a laboratory where I could continue my researches . . . my hopes have been dashed . . .'.[9] He is, therefore, he says, taking the opportunity to publish what he can salvage from the wreck of his career so far, with apologies for its incomplete nature; it was almost as if he did not expect to live long. As his obituary says, 'Like Jean-Jacques Rousseau, he fancied he had more enemies than he really had.'[10]

In 1838, on the recommendation of Dumas, Laurent was appointed professor at Bordeaux. The facilities there were almost non-existent. The laboratory had no ventilation, light or heat and Laurent had to do most of his reactions at home. Despite all this, he was an experimenter of unparalleled skill, able to produce meaningful results with only the cheapest materials and equipment. Some attributed this penury to his excessively humble spirit, but humility is not the first epithet that comes to mind in perusing his letters; quiet determination and fighting spirit stand out. He had, however, married in 1839 and at about the same time had lost 10,000 francs in a commercial venture. It is not difficult to imagine how tarring him with the brush of church mouse might have been an excellent way for Paris rivals to sideline him as someone ineffectual. His controversial theories also alienated him from many in the metropolitan establishment.

It is true that Laurent was disorganized. His many ideas were often expressed in an idiosyncratic way and could be ignored by others until they were shown to be valid, when someone else invariably took the credit. When it came to nomenclature, he was his own worst enemy in inventing bizarre new systems; his name for aluminium alum was 'atolan-telmin-ojafin-weso'. No wonder that, when he published his theory of organic chemistry consisting of fifteen propositions, in 1836, Berzelius remarked: 'I think it useless that my reports shall deal in future with such theories.' But sometimes such eccentrics as Laurent are to be cherished. Deep in this arcane morass was one of the most important concepts in chemistry.

Charles Frédéric Gerhardt, eight years Laurent's junior, was of Swiss descent and brought up in Alsace; his family were brewers. He initially trained as a chemist on the initiative of his father, who had invested a considerable sum of money in a venture to build a lead carbonate factory; the disappearance of a business partner had left him saddled with an enterprise of which he knew nothing. After qualifying, Gerhardt did obey his father's wishes for a time, but, like Laurent, he found factory work tedious and noxious; he rebelled and joined the 13th Regiment of cavalry. Predictably, this soon produced an overwhelming desire to escape, and he was bought out for 2,000 francs lent by a friend. This benefactor appears to have been Liebig.

After leaving the army, Gerhardt went straight to Giessen, where he worked with Liebig for a few months, and undertook some translations of his work into elegant French; brought up in Alsace, Gerhardt was bilingual. Why he did not stay longer is uncertain. His testimonial from Liebig says that Gerhardt would be a promising worker once he had restrained his impetuosity and recognized that successful work required steadiness and application. His biographers say that Gerhardt was unhappy that Liebig wanted to marry him off to his daughter, and that he found Liebig domineering.

By the time he was twenty-two, Gerhardt, with 200 francs given to him by his brother-in-law, had set out for Paris, walking part of the way. There he was initially well received as the translator of Liebig's works, and became Dumas's assistant, but, like Laurent, he did not find the atmosphere congenial. The argumentative Gerhardt would keep advancing theories, which Liebig wrote to warn would be professional suicide: 'The Academy has always reserved to itself the right to make scientific laws, and it considers those who would take over this function as thieves and assassins; a young man who forces them to teach these laws to him cannot expect any advancement . . . for the love of God, do not write any theories except in the German journals!'[11]

What good was it for Liebig to advise him that the way to advance his career was to keep coming up with interesting and uncontroversial new research results, he wrote back, when Dumas never gave him any decent laboratory facilities? How was he to survive with neither the practical means to get his head down nor the ability to intrigue and dissimulate?

But, for the moment, he was astounded to be appointed, like Laurent, to a provincial professorship, in Gerhardt's case at Montpellier. But, like Laurent, on arrival there he was in no time dissatisfied with the level of funding. It was supposed to be the second centre of medicine in France, but his colleagues did nothing but earn money on the side. He had to clean his own apparatus, the climate did not agree with him and the town and the university were filthy. The locals of the Midi were unsociable, 'barricading themselves inside the house in case anyone should make off with their wife or daughter'.[12] He suspected Dumas of having sent him to Montpellier to get rid of him; worse, he wrote to Liebig and told him so. Ironically, his real or imagined exile had the effect, because of the lack of facilities, of forcing him to concentrate on reading the literature and on theorizing against all advice, rather than carrying out uncontroversial practical researches.

In 1844, despite his continuing lack of funds, he married Jane, the daughter of John Richard Gordon, a Scottish veteran of the Peninsular Wars who had retired to Montpellier after killing his fiancée's brother in a duel.[13] Liebig was gratified; marriage would cool Gerhardt's hot head, and he was to be congratulated on netting 'une anglaise' and thus drawing such an enviable number in life's lottery. But, if he thought that Gerhardt would settle down to a life of placidity, he was gravely mistaken. The subsequent years saw the emergence of Gerhardt's ideas about the molecule against a background of continuous strife – personal, professional and political.

Half a century on from 1792, France remained deeply split between the heirs of the Revolution and the Bourbonist counter-revolutionary establishment. It was inevitable that the sympathies of both Gerhardt and Laurent should be against the reactionaries personified by Dumas. When the 1848 Revolution broke out, there were student disturbances in Montpellier. Gerhardt left hotfoot on the midnight stagecoach to Paris, there to lobby ministers, wheel and deal and do all that he could to dislodge the 'Cumulards', the holders of multiple academic posts, and thus stir up their undying antipathy. He wrote to his shocked Scottish mother-in-law in 1850: 'There are but two parties in France, the men of the past and those of the future. The latter have been called under the restoration, liberals; under Louis Philippe republicans, and today they are called socialists. The liberals have been guillotined, the republicans shot and logically the same treatment will be accorded to the socialists . . . it is the battle of the weak against the strong, of justice against brutal force . . .'.[14] Like all such revolutionaries in a hurry, he had difficulty in distinguishing the general good from his own personal interests. Hanging around the fringes of the Paris insurrections of 1851–2, he was at one point arrested but released. On the afternoon of 4 December 1851 he went with some friends, including the chemist Sabatier, to the Porte Saint-Denis to find out the state of play at the barricade that had been erected there; they were caught up in the fusillade at the rue Richelieu and had to take cover for several hours in a doorway. They eventually reached Gerhardt's apartment with difficulty, as all the Seine bridges were guarded; Sabatier, an activist, hid out there for several days avoiding the police, and eventually made good his escape to Belgium.[15]

The new ways of thinking about organic chemistry that were to topple Berzelius's electrically based dualistic theory actually began with Dumas. The researches providing the first clues had to do with substitution reactions involving chlorine.

One day in the early 1830s, a soirée was held at the Tuileries gardens, at which the guests were distressed by the acrid and irritating fumes given off by the candles being used. From this minor inconvenience, a new theory of chemistry would eventually emerge. Investigation of the matter was passed to Brongniart, who in turn passed the buck to Dumas, his son-in-law.

Dumas collected the fumes and found that they were hydrochloric acid (hydrogen chloride). He found out that chlorine had been used to bleach the wax used in the candles. It looked as if the bleaching process had somehow modified the wax so that it now contained chlorine. He carried out some experiments concerning the action of the halogen on various other substances containing carbon and hydrogen only, such as oil of turpentine, and found that chlorine could replace hydrogen in them without any major discontinuity in properties.[16] The reactions that the wax or turpentine seemed to be undergoing can be written as follows:

$$[\text{HYDROCARBON}]\text{--H} + 2\text{Cl} \rightarrow [\text{HYDROCARBON}]\text{--Cl} + \text{H-Cl}$$

But, according to Berzelius's theories, this could not be possible. The electronegative atom chlorine could not replace the electropositive hydrogen, or if it did there would be a profound change of properties. For example, it was impossible to replace the chlorine in salt (NaCl) by hydrogen. (Sodium hydride, NaH, was unknown; it was not made until 1902 and is highly reactive, decomposed violently by water and quite unlike sodium chloride.) In another experiment, by chlorination of alcohol Dumas obtained a substance that he called chloral. This retained the oxygen content of the alcohol and, although different in its properties, was clearly a substance of the same *kind*, and not the profoundly different entity that the dualistic theory would have required.

Dumas summarized his findings in three rules governing these substitution reactions, one of which was incorrect, and applied the rules to a range of known reactions. 'Organic chemistry, so rich in detailed facts, lacks general rules,' he wrote. 'Most of those given in treatises of chemistry are pure illusions.'[17] One reason why this was undoubtedly the case continued to be the gross errors in assigning molecular formulae. Dumas's formula for alcohol, for example, was $C^8H^{12}O^2$, which, struggling to make sense of in ways that might explain its reactions, he wrote as $(C^8H^8 + H^4O^2)$. The correct formula is C_2H_6O.

Dumas could easily find other examples of substitution reactions. By chlorinating acetic acid, he obtained a substance, trichloracetic acid, containing three equivalents of chlorine, which resembled acetic acid in its properties and reactions. Berzelius first of all disputed the fact that they were similar; then, after Dumas had produced further evidence that they were, flatly stated that the whole scenario was impossible. This concept of substitution will have an 'influence nuisible' on the progress of science, he wrote. 'An element so eminently electronegative as chlorine can never enter an organic radical; the idea is contrary to the first principles of chemistry.'[18]

Laurent, Dumas's assistant at the time, took the experiments further with extensive work on the chlorination of naphthalene (a close relative of benzene), at first in collaboration with the senior chemist. In an extraordinarily painstaking and lengthy series of investigations, he isolated and purified numerous chlorination products of the hydrocarbon. This work culminated in his lengthy 1836 paper containing his fifteen propositions,[19] which Berzelius dismissed as rubbish: 'A kind of legislation, which, like all hasty legislation, appears in numerous regulations.'[20]

Like all attempts at this time, Laurent's propositions contain many misapprehensions. His formulae were all over the place, with naphthalene shown as containing forty carbon atoms whereas it only contains ten; not only was he using the incorrect atomic weight of 6 for carbon, but he was sticking to the incorrect '4-volume' formulae. These owed their genesis to the now definitely outmoded formula HO for water, but the link had been so mangled in Berzelius's thought machine that it was not clear to the majority of chemists. There were far too many atoms lurking unaccounted for in most of these formulae.

The most significant of Laurent's propositions was the fifth: '*All organic compounds are derived from a hydrocarbon, a fundamental radical*, which often does not exist in its compounds but which may be represented by a derived radical containing the same number of equivalents.'[21]

The terminology is confused, but Laurent is groping towards one of the most important concepts. Organic chemistry is largely the chemistry of functional groups (a term not yet invented) attached to a relatively inert hydrocarbon skeleton. The properties and reactions of a compound depend on the functional group(s) it contains. Thus both acetic acid and trichloracetic acid are carboxylic acids, and their properties are similar. Laurent's use of the word 'radical' is confusing and incorrect, a point that Liebig seized on; a hydrocarbon is not a radical, but, elsewhere, Laurent uses the much better term 'nucleus' in place of it. He says that chlorine in the substitution products can 'play the part of' hydrogen without a fundamental alteration of properties.

Laurent was later disgusted to find that his theories had first of all been dismissed as absurd, and then anything useful in them had been attributed to Dumas. 'Justice for all, for the great and the small, is all I demand . . . If my theory fails, I am responsible; if it succeeds, it is the work of someone else.'[22] The silver-tongued Dumas was not averse to encouraging such legerdemain. On the whole though, most observers at first agreed with Berzelius, who called this new attempt 'thick-fog chemistry'. Laurent complained that his contributions were given less than their fair due because he had been 'exiled' in Bordeaux where people came to his lectures only to find out how they could adulterate their wine better. Liebig at first thought them completely unfounded, the first researches of a beginner, the whole theory an arbitrary one developed in complete ignorance of the fundamental principles of science. Laurent was mortified that Liebig had at first ridiculed him – though in 1843, when he visited Giessen, Liebig was kindness itself and let Laurent lecture to the students on his theories. But by 1843 Liebig was coming round to the view that there was something in Laurent's ideas after all.

In 1840 Dumas developed his ideas into what became known as the 'type theory', something highly incomplete in his hands, but which would undergo further development by Laurent and Gerhardt. In an organic compound, he said, all the elements can be successively displaced and replaced by others. He criticized the electrochemical (dualistic) theory. Although electrical forces certainly had an effect on chemistry, progress was hamstrung by the theory's requirement to identify the positive and negative parts of every compound: 'No view was ever more fitted to retard the progress of organic chemistry.'

Dumas included a caveat that both the electrochemical and his type theories could, if pushed to the extreme, result in absurdity. This did not prevent the appearance of a lampoon on his theory appearing over the signature of 'S.C.H. Windler' ('Swindler'), published in Liebig's *Annalen*. This might lead us to guess that it was penned by an intimate of his, and the author was indeed later identified as none other than Wöhler. Writing significantly in French from Paris, 'Windler' informed the readers of the journal that he had verified Dumas's substitution theory in a most remarkable and unsuspected manner. Manganous acetate $MnC_4H_6O_4$ (modern formula) was treated successively by chlorine in various ways over a period of four years. It gave first a 'violet-yellow crystalline substance' and then, via further intermediate stages, pale-yellow crystals $ClCl_4Cl_6Cl_4$. In this substance, every atom of the starting material had been replaced by chlorine, with no appreciable change in properties! Furthermore, a footnote informs us, the technology was already being applied to cloth, and bleached fabrics consisting entirely of spun chlorine were currently on sale in London and much in demand by hospitals for the manufacture of nightcaps and underpants.[23]

In the face of this ridicule, Dumas backpedalled furiously: 'I have never said that the new body formed by substitution has the same radical, the same rational formula as the first. I have said exactly the opposite on a hundred occasions. M. Berzelius has assigned to me an opinion that is not mine.'[24] He was to regret this, because by disowning Laurent's fifth proposition he was disowning his share of the credit for a vital insight.

The agreeable if sardonic Wöhler was always something of a spectator to the titanic clashes of personality that these shifting tides of opinion caused. Liebig began to come round to the type theory, which caused a final break between him and Berzelius. In 1845, three years before the latter's death, he visited Germany for the last time, but made a long detour to avoid having to pass through Giessen. As Partington says, 'The history of organic chemistry about 1840 presents a distressing picture of animosities between Berzelius, Liebig and Dumas; when they combined forces and turned on Laurent and Gerhardt the effect was even more deplorable.'[25]

Poor Laurent was taken to task again and again for some real or imagined error. He prepares a new compound and remarks on his surprise at its highly irritant and pungent odour. Why, replies Berzelius, if only he were not so blinded by his unhappy substitution theory, he would have known that this was completely expected; my annual reports to the Stockholm Academy could furnish a mass of examples.[26] In 1845 Laurent complained with justification that, in Berzelius's annual review of the most important chemistry of the last year, the 'crafty old so-and-so' (*rusé matois*) had made no mention whatsoever of Laurent's vital work on naphthalene.[27] Laurent was always more renowned outside France than in it, especially in Britain. William Gregory, Professor of Chemistry at Edinburgh, wrote: 'We read these amazing works and we see without difficulty that they are the works of the foremost chemist of the day whether he be in Paris or elsewhere. Those who know France a little, and what Paris is to France, are beginning to be astounded that M. Laurent remains at Bordeaux, neglected, almost misunderstood, by his compatriots.'[28]

The ultimate indignity came in 1850, when the chair at the Collège de France became vacant. Gerhardt declined to stand, because he thought he was so unpopular that he stood no chance. Laurent was persuaded to stand. Although the majority of the professors recommended Laurent, the Academy voted by thirty-five votes to eleven to award the chair to Balard, a third-rater who already had two other chairs and did no research. (Balard's only useful achievement had been discovering the element bromine in 1826, purely by chance; an unkind wit has said that 'bromine discovered Balard'. Gerhardt said that most people would be surprised to discover that Balard had not died years before.) On the very day, Monday, 30 December 1850, that the Academy met to consider the appointment, one of Balard's supporters, Fremy (whom Gerhardt referred to as a 'Limus philosopher'), read a paper to the Academy in which he laid into Laurent and Gerhardt for getting some of the chemistry of tartaric acid wrong.[29] The work is of 'incontestable importance', he says, and the 'well-known ability' of the chemists siding with him contrasts with the two meddlers. Furthermore, the intervention of an 'illustrious physician' into a purely chemical matter has made it of signal importance that the resolution of the problem should be in the hands of responsible persons who will not bring the whole science of chemistry into disrepute. He does not wish to 'abuse the time of the Academy' but he cannot forbear to express astonishment that the two failed to observe certain facts, and so on and so forth. The tone is egregious throughout. Laurent, who seems already to have been suffering from tuberculosis no doubt exacerbated by the inhalation of lung-weakening fumes, suffered a collapse and was unable to do any more laboratory work until his eventual death something over two years later. He was forty-six.

For the majority of chemists the shift away from Berzelius's electrochemical theories was slow. Berzelius himself never really accepted organic substitution reactions until his death in 1848. His formulations of organic compounds became more and more convoluted and he was forced into devising formulae for imaginary entities. He introduced the concept of 'copulae' (a term in fact picked up from Gerhardt, but given whole new meanings that Gerhardt had never intended) to try to give his theories new life, but the copulae theory was a product of his own imagination that no one else fully understood. Unfortunately, in defence of his ideas, Berzelius was dogmatic and navigated to the sound of an inner voice. As he had said in 1839, 'When I see an incorrect theoretical representation, I feel, even when the correct one is unknown to me and *when I cannot make it clear to myself*, that it is wrong . . .'.[30] Laurent said, 'What then is a copula? . . . an imaginary body, the presence of which disguises all the chemical properties of the compound with which it is united. Is not the dishonesty flagrant? What have I to do with your copulae and your radicals and all your castles of cards?'[31]

Bunsen's fearfully dangerous researches on arsenic compounds around 1840 appeared in some ways to support the copulae theory and Berzelius was jubilant: 'This is a triumphal chariot which has smashed the ramshackle barricades of Dumas,' he wrote.[32] In 1844 a German textbook could still state that the electrochemical theory was accepted by most chemists. The 1847 edition of the same book says that it is not clear what the outcome of new theories would be; no mention is made of Laurent or Gerhardt. Most chemists were bemused by the sheer complexity of the clash of ideas.

Only one photograph of Laurent exists; he leans on his stick like a hermit or monk, the latter impression accentuated by some kind of waistcoat that he is wearing, the pattern on which makes him look as if he has a large crucifix round his neck. Several extant photographs of Gerhardt, on the other hand, show someone saturnine, his dark-shadowed eyes perhaps presaging his early death. We can only speculate what the two might have achieved had they survived into the 1860s.

By 1845 Gerhardt was writing to his friend Cahours: 'You really must make the acquaintance of Laurent. One must know him to appreciate him. . . . I assure you, he is the best pal [*garçon*] in the world, a bit touchy on questions of priority, but think how often he has been done wrong . . . what an astonishing brain, and what a crammer [*piocheur*]. . . . I have never come across a man with such an abundance of ideas, his letters to me are something precious, I study them like a disciple.'[33] The first surviving correspondence between Laurent and Gerhardt is dated July 1844 when their relationship was still rather formal. Before long, it became intimate and conspiratorial. Large parts deal with trying to correct the analyses and formulae of organic compounds, arising from their intimate knowledge of the literature. With the benefit of hindsight, like most chemists of that era they spread their net too widely. Gerhardt spent much time recalculating the formulae of various alkaloids, the true chemical nature of which was not to become clear until well into the following century, and Laurent was never going to discover anything of immediate value from his studies of the formidably complex element tungsten.

But there was always room for complaints, mutual exhortations and lampoons, even cartoons by Laurent, such as one in which he shows a chemical reaction as two rats eating each other head-on. It was the two of them in comradeship against the world. An 1846 letter from Laurent includes an ironic coat-of arms that he has designed for the newly created Baron Liebig. One quartering of the shield shows Liebig's famous analytical apparatus of glass bulbs for potash. The other segments represent three of the Baron's recent mistakes; one of them a melon drawn to represent a substance called mellon, concerning which Liebig had engaged in a quite unnecessarily vitriolic dispute with Gerhardt until the latter had been proved right.[34] Elsewhere Laurent takes Gerhardt to task for writing an indiscreet letter to Quesneville in which he describes all the Parisian chemists as 'cretins and ignoramuses', a letter that Quesneville was gleefully hawking all round the city.[35]

Another theme is their often-repeated desire to be together. The logistical problems were not trivial. In July 1845 Gerhardt travelled from Montpellier to Bordeaux, by coach as far as Sète, then by boat, first through the marshes of Thau, where he was seasick, then along the Canal du Midi, where he got sunstroke; a journey of more than three days. The pair then travelled from Bordeaux to Paris and spent nearly a month together at the Hôtel de Saxe in the Latin quarter. When travel was so slow and expensive, the two had a perfectly rational reason for their extended meeting together. But glimpses of a much more intimate relationship come in the letter in which Laurent addresses Gerhardt as 'Ma chère Victoria' and the vignette that he drew two years later showing him as the queen of the same name (see Plate 11). 'I am disheartened to be so alone,' Gerhardt wrote to Laurent on 26 February 1846, and in the margin adds: 'Think of the arrangements to be made for us to be together again; lodgings, laboratory. I am firmly decided.'[36] And in other letters, a few weeks later: 'Do you think always of Paris? . . . I fear to commit you to coming to Paris, but I desire it with all my heart. . . . I can never be happy except when I am with you.'

On his deathbed, Gerhardt, self-pityingly, wrote: 'In fifty years time, it will be seen that I have really done something.'[37] Whether this is a piece of justifiable hyperbole that we can allow a dying man is a matter of opinion. In general, history bore him out. There is no doubt that his work, added to that of Dumas and Laurent, took chemistry forward several massive steps, but the exact apportionment of the credit is another matter.

In a situation where different chemists were using different ways of arriving at the formulae, and where the majority of those formulae were grossly inaccurate, Gerhardt used a different approach. He tabulated the known compounds, first by the number of carbon atoms that they appeared to contain, and secondly according to analogies in their properties, which pointed to the presence of the same functional group. Ranking by the apparent carbon number avoided the difficulty that the exact number of carbons thought to be in most compounds was wrong, and skirted round the atomic weight problem. The placing would be unaffected whether the difference between adjoining members was C_2H_4 (as most thought) or CH_2 (the correct answer). Gerhardt introduced the important concept of *homology* – that is, compounds with similar properties differing in the number of carbons in their formula, for example, methanol (methyl alcohol) and ethanol (ethyl alcohol or ordinary alcohol). Both are alcohols; and both contain the functional alcohol group –O–H, attached to different carbon frameworks.

As Gerhard said, it was like laying out a deck of cards in rows and columns according to rank and suit. One column contained the various acids that were then known: acetic, propionic, butyric (isolated from butter in 1817 by Chevreuil), valeric (which the same chemist isolated first from porpoise fat), benzoic and so on. Another column contained the hydrocarbons, another the alcohols, and so on across the table. Compounds that could be easily interrelated by simple reactions without gain or loss of carbon he placed in the same row: thus alcohol, acetic acid, ethylamine, ethyl chloride. There had been earlier attempts to find regularities in the properties of compounds sorted by formula, but the attempts had neglected to separate the series of compounds of different chemical type (functional group) as Gerhardt had done. A compound in which an oxygen atom, say, is hidden inside the molecule (an ether) is quite different from one in which it is on the exterior in the form of an –O–H group (an alcohol).

Using modern formulae, we can easily show how Gerhardt's rows and columns make perfect sense. It is, however, vital when looking at this table to remember, first, that Gerhardt had no idea of

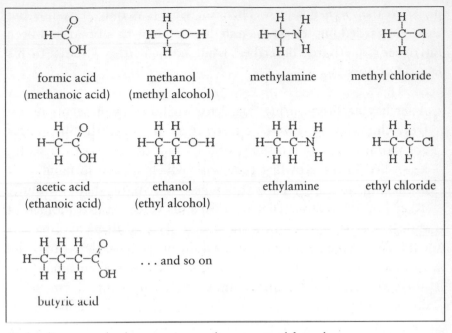

Gerhard's rows and columns, using modern structural formulae.

molecular structure, and did not even believe that such a thing existed, and secondly that his molecular formulae were incorrect. In the light of this, to come up with his table of homologies was a major achievement, and one that could be accomplished only by someone with Gerhardt's encylopaedic knowledge.

Immediately he had done this, regularities began to appear. Dumas said it knitted compounds together by 'quite unsuspected threads'. Whether Dumas ever tried to claim all the credit is not clear, but Gerhard thought he did, merely by not contradicting others when they gave him the credit.[38] In the hands of others, such as the English chemists of the 1850s, these ideas would come to dominate thinking about the molecule.

On 5 September 1842 Gerhardt presented to the Academy his paper *Researches on the Classification of Organic Substances*, which he had written in exile at Montpellier. This paper unleashed, in the words of his biographers, a 'veritable tempest'.[39] Thenard thought the tone impertinent, Dumas tried to bag all the credit, and Regnault was incandescent that some of his own formulae had been shown to be false. Soon afterwards Gerhardt called to present a copy of his paper to the *éminence grise* Thenard, who was only sixty-five but clearly regarded himself as the antediluvian heir to all that had been Lavoisier's. Gerhardt described what happened in a letter to his friend Cahours:

> After having been on my guard for an hour awaiting his return from the country, I was at last taken in to see the illustrious collaborator of Gay-Lussac. I offered him the copy of my work; he replied in the growling tone with which we are so familiar, '*I am not at all happy with this mémoire*'. Astonished, I timidly asked him the reason. Thenard raised his voice. At that moment, I thought he was going to eat me . . . he shouted that Lavoisier, the most distinguished of chemists, would never have dared to write that such and such a thing is false, [an educated Frenchman] did not write like that. 'Unfortunately', I told him, 'I am not a courtesan; when I have a conviction, I express it straight out, but . . .'. Thenard would not let me finish; he replied, 'I do not like courtesans, but I tell you that five or six members of the Academy are absolutely of the same opinion as me; are you convinced?, I, not hearing properly or misunderstanding the last phrase, said 'No.' Thenard sprang to his feet, '*Adieu, Monsieur! Adieu Monsieur!*', he said at least ten times, and without letting me collect my thoughts propelled me out of the door in the presence of his servant. I expected to receive a kick up the ——, and was amazed that one was not administered.[40]

The biographers of Gerhardt, who included his son, were in no doubt that he was ill treated by the establishment. His battle for truth roused against him a coalition of jealous and incompetent placemen ('incapables en place') mounting 'attaques perfides', and his whole life was a sacrifice to science.[41] Charles Gerhardt though is not very likeable. One comes to feel that Dumas and Liebig had much on their side when they fended off this aggressive parvenu. In 1846 Liebig wrote to Laurent, 'I believe you to be a man of perfect loyalty and honesty who . . . has linked his destiny with that of a man without character and without morality.'

It is possible to get a sense of Gerhardt's morality, and especially of his attitude towards the intellectual property rights of deceased persons, from a letter of his to Chancel in November 1852. This describes the fate of the French translation of Berzelius's book. (Berzelius had died in 1848.)

> Great news [*Grrrande nouvelle*]! I have sold my Organic Chemistry for the bagatelle of ten thousand francs [*balles*] to the firm Didot. Here is how it was arranged; Didot *Fils* visited me about a fortnight ago and proposed to cease the Berzelius, the author having left organic chemistry unfinished; he proposed leaving off after two volumes. I replied that I was not interested in patching up this work, as the section on organic chemistry was very bad, and I had nearly finished a book of my own on the subject . . . several days later, I made him the following proposition; publish my four volumes, under the title 'Berzelius's Treatise on Organic Chemistry, entirely rewritten and augmented [*enrichi*] with recent discoveries by Ch. Gerhardt' and on the cover 'Ch. Gerhardt, Berzelius's Treatise on Organic Chemistry' . . . Laurent thinks that if poor Berzelius had suspected what I would one day do in his name, he would have died a week earlier.[42]

Liebig, in a dispute over the formulae of certain compounds related to urea (in which Gerhardt was right and he was wrong), had penned a vitriolic polemic against Gerhardt entitled 'Herr Gerhardt and Organic Chemistry'. He compared Gerhardt to a highwayman (*Strassenrauber*), and, even allowing for Liebig's hot temper, may not have been so wrong. (The English chemist Odling, not to be left out, said, 'They called Gerhardt a brigand, and he certainly looked like one.')

In 1867–8 another Alsatian chemist and assistant to Dumas, Charles Adolph Wurtz, wrote about Laurent and Gerhardt, and their names became indelibly linked together; 'The great figure of Gerhardt will not be separated from Laurent; their work was collective, their talent complementary, their influence reciprocal.'[43] Wurtz, an influential figure, had been at school with Gerhardt, but the reason for his writing this was to resurrect the reputation of *Laurent*, who had become almost forgotten. Why?

In the last months before Laurent's death, the two became estranged. Gerhardt was not invited to the funeral and says that he learnt about it only a few hours before 'by a simple announcement'. The correspondence between Gerhardt and other acquaintances at this period goes quiet. Chancel fails to reply to his letters, and, although Gerhardt bluffs it out, one senses that his ingenuousness is hiding something. No one showed him the manuscript of Laurent's *Méthode de chimie*, which went to press immediately after the funeral. No obituary of Laurent was written by his fellow countrymen. One appeared in the United States and one in England; those published in France were translations of the latter, by Williamson. This makes only one, unimportant, reference to Gerhardt, and says that '[Laurent] had no enemies; he sometimes met with ingratitude, even from those on whom he had conferred the greatest benefits – such ingratitude, in fact, embittered his last moments'.[44] Laurent had friends across the Channel who thought that Gerhardt had laid claim not only to their joint ideas, but to others that Laurent had passed to him as he lay dying. Gerhardt may have tragically betrayed his friend.[45]

Gerhardt survived Laurent by only three years, struck down with peritonitis in August 1856 on leaving church, and was buried on what would have been his fortieth birthday. Chemistry moved on. Gerhardt's classification and the radical system would triumph over Berzelius's electrochemical ideas. The concept of functional groups and carbon frameworks was being born, and awaited only a resolution of the atomic-weight question, but Gerhardt did not live to see the results.

The chapel at which he attended his final mass was destroyed by bombardment fourteen years later during the Franco-Prussian War. The bronze statue of Auguste Laurent by Péchiné that in 1903 was erected in the town square at Langres where he was born was taken away and melted down by the Germans during the Second World War.

ELEVEN

A Cross-Channel Excursion

CAROLINE. *To confess the truth, Mrs. B., I am not disposed to form a very favourable idea of chemistry, nor do I expect to obtain much entertainment from it. I prefer those sciences that exhibit nature on a grand scale, to those which are confined to the minutiae of petty details. Can the studies which we have lately pursued, the general properties of matter, or the revolutions of the heavenly bodies, be compared to the mixing up of a few insignificant drugs?*

Mrs Marcet, *Conversations on Chemistry, in which the Elements of that Science are Familiarly Explained and Illustrated by Experiments*, 2nd edn (1807)

At one time Britain had been in a ferment about the new science of chemistry. The glamorous Davy and his Royal Institution lectures had a lot to do with this. But Davy had died in 1829, aged only fifty, and during the next couple of decades the subject lost its glamour. It became a byword for tedium and was looked down upon by many 'real' scientists as a second-class pursuit, hardly a science at all. Most came to agree with Sir John Herschel, the astronomer and author of the influential *Discourse on Natural Philosophy*, who rather patronizingly thought that chemistry would very soon be reduced to the status of a deductive mathematical science. Its laws were 'For the most part of that generally intelligible and readily applicable kind, which demand no intense concentration of thought and lead to no profound mathematical researches'.[1] Davy's most able successor in England had been Michael Faraday, another sceptic, but after about 1825, when he discovered benzene, his researches were mostly in electricity, not in pure chemistry. Neither had the study of chemistry theory produced any useful practical discoveries. The attitude towards it in Britain was mostly one of 'sceptical empiricism'. Chemistry was a useful technology, but why fiddle about with the theory when Davy, who did not believe in atoms, could come up with useful inventions like the miner's lamp, and electrochemically improving the resistance of His Majesty's ships to corrosion?

In part, this was the besetting sin of insularity. An outsider's view was given by Berzelius, *c.* 1830. Laid up in bed for several days, he passed the time reading some foreign chemistry texts, including Turner's *Elements of Chemistry*, published in 1827. 'What damned nonsense,' the irascible and gout-ridden Swede reported.

> Written in an impertinent style, oblivious to everything which has not been done in England by Englishmen . . . the chemists of England live in their own world . . . [Smithson] Tennant [the discoverer of iridium and osmium] told me that he intended to put out a history of chemistry's last fifty years, but I have entreated him to forget about it, since when I questioned him a little he showed himself to be consistently misinformed. . . . The commercial instincts of the English nation find expression even among their learned. There is a great deal of litigation here about priority in the most petty matters, and despite the fact that they get on fairly well with one another, one can regard them as puppies who stand and snarl over their bones, from which the meat has been gnawed on the Continent. Davy growls and shows his teeth and Wollaston lies carelessly beside his.[2]

It was the foreigner Liebig who did as much as anyone to raise the profile of chemistry in Britain again. Many of the new generation of British chemists had been his pupils; British students at Giessen were the largest foreign contingent. He was a popular and welcome guest in England both in scientific circles and, when he became increasingly involved in commerce through his meat extract and other ventures, with the general public. This reflected the climate at national level. Relations between Britain and Germany were generally cordial. Later, at the time of the Franco-Prussian War, public sympathy in Britain was overwhelmingly with the Prussians.

In 1837 and 1842 Liebig made tours of the British Isles. On his first visit he travelled from Manchester to Liverpool on the railway in thirty-seven minutes, and was mightily impressed by the two miles of ships lying in Liverpool docks. He met Faraday, who lionized him at a banquet. He became an anglophile, later read Macaulay's *History of England*, and came to the conclusion that the island people, with their respect for the law and free political system, had something the Germans lacked.[3] But he had not been very impressed with the scientific climate: 'I have just spent several months in England, where I saw a lot and learnt little,' he wrote to Berzelius at the time of his second visit. 'England is not the land of science; it can show only a widespread dilettantism. The chemists are ashamed to call themselves chemists, because the despised apothecaries have run off with the name.' He tried to impress on the British chemists the coming importance of organic chemistry in medicine, and regretted the fact that it seemed to have no roots in England.

Nevertheless, in the period leading up to the untimely deaths of Laurent and Gerhardt, and for a little while afterwards, the main locus of thought about the molecule was, for a brief few years, and rather surprisingly, Britain.

Prince Albert had been a student at the University of Bonn. When he and Queen Victoria visited the town in the early 1840s as part of a Beethoven festival, he took her to see the rooms where he had once lodged. They found them converted into a laboratory where August Wilhelm Hofmann, the son of the architect of Liebig's Giessen laboratory and his favourite pupil, was working as assistant professor. They asked him to show them some experiments, which the Queen found impressive. As an eventual result, a new college was founded at 16 Hanover Square, London, and Hofmann was appointed in charge. The story sounds too good to be true; there had obviously been some string pulling behind the scenes preceding Her Majesty's visit. Whatever the truth of the matter, the Royal College of Chemistry became the first institution in London to have a research laboratory on the Liebig model, and chemistry in Britain received a shot in the arm. It was from this college that William Perkin, one of Hofmann's students, decamped in 1856 to found the British synthetic dye industry.[4] Hofmann became related twice over to Liebig; he was married three times, and both his first and third wives were Liebig's nieces by marriage.

London especially became a centre of debate amongst those who knew each other from the Giessen days, and others. After the building of the first generation of railways and the foundation of the Chemical Society of London (in 1841), the capital over the next few years could serve as a meeting place for Alexander Williamson (b. 1824), Edward Frankland (b. 1825) and William Odling (b. 1829) among others. There was also the Scot, Archibald Scott Couper (b. 1831), although he spent most of his time on the Continent.

British chemists of the period were a mixed bunch, none of them with the drive and personality of a Davy or a Faraday. Many were introspective academics, reflecting the nature of a subject that was to all intents and purposes still in the doldrums. Perhaps this did have the advantage that they were thinkers at a time when chemistry had too many observations and not enough good theories. But, ever since England had been described by Liebig in 1837 as being the land of dilettantism, and before, the climate had been an amateurish one. In retrospect it says a lot for the intellectual qualities of this small band that they were able to achieve so much in such a relatively sterile environment. But the overall tone is still not truly professional; Couper, for example, began by studying philosophy at Edinburgh, and took up chemistry only after he had moved to Berlin in 1855. Even the Royal College was atrociously short of funds. After only three years of operation it had to move to smaller premises at 299 Oxford Street; Hofmann had to give up his free accommodation, and the college secretary, who refused to move, had to be evicted by force.

Only one of the four mentioned above, Odling, had a life relatively free from woe, and even he, much later, met an unfortunate end. He was run over and killed by a cyclist in Oxford in 1921 after retiring from his post of professor. (Remarkably, his opposite number, Liveing, Professor of Chemistry at Cambridge, suffered exactly the same fate three years later.) Williamson battled against physical disability; he was paralysed in one arm and nearly blind in one eye. After about 1855 he more or less dropped chemistry and took up university committees and steam boilers. Poor Couper suffered a breakdown from stress soon after his important contribution, and never recovered.

Frankland was almost equally unfortunate, for he suffered from that most crippling of Victorian disabilities, illegitimacy, which blighted his career and personal life. In the opinion of many, he was one of the leading scientists of his era, his reputation unfairly neglected partly because of this stigma and partly because of his difficult personality.[5]

What was the concept that was struggling so painfully to be born in the 1850s and 1860s? It can be encapsulated in one word, although that word was not properly given birth until the end of this period: *valency*. The word in its final form first appeared in an essay by 'Sigma' (unidentified) in the *English Mechanic* magazine of November 1869.[6] Valency, expressed in its simplest form, is *the combining power of one atom for another*.

Although old dictionaries contain a word 'valency', derived from medieval Latin *valentia* meaning 'value', this usage had become obsolete long before the nineteenth century, and none of the chemists seems to have been aware of it. In the middle of the century, a confusion of words reflected the confusion of thought as the concept emerged. Eventually Hofmann, who was bilingual and proud of his command of English, rejected earlier terms and coined the English word *quantivalence* in 1865. This word crossed the Channel with him when he moved to Berlin, where they found it too long. It was shortened to the German *Valenz* and then came back to England as 'valency'.[7]

The simplest way of thinking about valency is as the number of 'hooks' that an atom has to join it to its neighbours. The uncertainties of the Berzelius era, still unresolved at the time of his death in 1848, concerned the relationship between atomic weights and equivalent weights. For simple binary compounds such as water, we can write a very simple equation:

$$\text{atomic weight} = \text{equivalent weight} \times \text{valency}.$$

The equivalent weight of oxygen in water is 8 (an established fact). If the atomic weight of oxygen is 8, then its valency, the number of hooks on each atom, is 1 and water is H–O; oxygen atoms are monogamous. If, as is actually the case, oxygen's atomic weight is 16, its valency is 2, and water is H–O–H; oxygen is bigamous, or bivalent.

By the 1850s water had been pretty firmly established as H_2O for two or three decades, but the situation regarding the key element carbon was still completely unresolved. Berzelius had started off with the correct atomic weight of 12, but then changed it to 6. Gerhardt had changed it back again, but it was still uncertain. If the atomic weight was uncertain, the valency was uncertain. No one knew how many hooks the carbon atom had, and, if they did not know this, they could not get round to devising schemes of how the atoms might be joined together. This is even assuming that there was such a thing as 'joining together', for many chemists, including Gerhardt, did not believe that a molecule could have such a thing as a structure. The accepted representation was still Berzelius's bracketed formulae, with their implication that the constituent atoms were held together in a kind of electric soup. And let us not forget that a substantial number of people still did not even believe in atoms.

The concept of valency also had to be unscrambled from rather confused ideas about chemical *affinity*, not the same thing. The two phenomena are distinct, but this was not clearly recognized until the 1860s. In his 1865 book, Hofmann pointed out the contrast between compounds of oxygen with hydrogen, and of oxygen with chlorine.[8] The affinity of oxygen for hydrogen is very great, whereas compounds between oxygen and chlorine are unstable and liable to decompose explosively. But only two atoms of hydrogen combine with one of oxygen, whilst, in the highest chlorine oxide, no less than seven atoms of oxygen combine with two of chlorine; the valency of chlorine in this compound is 7. So affinity and valency are not the same thing. An atom can have powerful urges but be satisfied with a single partner, like Queen Victoria, or polygamous but sluggish, like her uncle George IV.

Confusion also arose from the difference between typical inorganic compounds and typical organic ones. Inorganic chemistry as a formal definition means the chemistry of all the elements except carbon. Organic chemistry is the chemistry of carbon compounds, although, of course, all organic compounds contain other elements too. Many elements show wide variations in their valency, or combining power. This is particularly true of some metals, such as iron, manganese and Vauquelin's chromium. Manganese ('the chameleon element', renowned for the way its different compounds

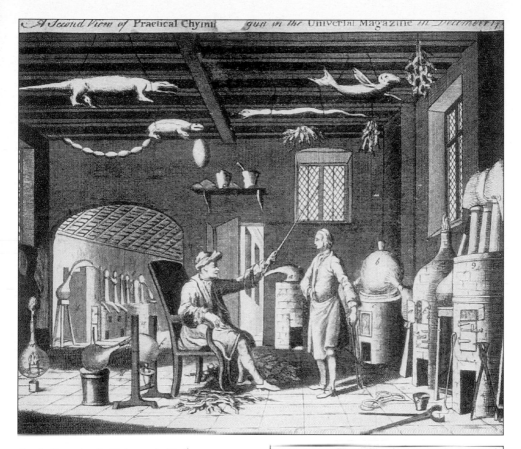

1. A Practical Chimyst's Laboratory, 1747.

2. Engraving of a Ceroxylon palm. Von Humboldt and Bonpland illustrated it and delivered the wax to Vauquelin, who analysed it, after a fashion.

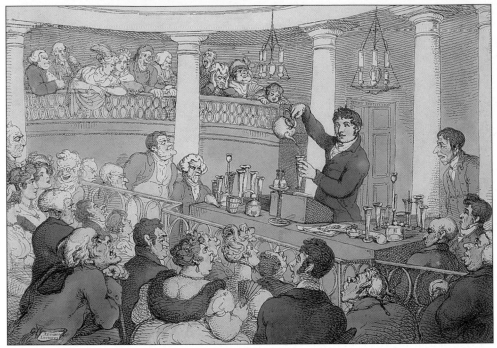

3. The Glamorous Davy, lecturing at the Surrey Institute, 1810. By Thomas Rowlandson. This is considered an accurate depiction, showing the popularity of his lectures.

4. Liebig's laboratory at Giessen, from the engraving by Trautschold.

6. Eilhard Mitscherlich, the 'Prince of Prussian Chemistry'. Hypersensitive and distressed by his arguments with Liebig and others. His great-grandson was the Freudian psychoanalyst who mounted a character assassination on Kekulé in the 1960s.

5. J.J. Berzelius at 65. Gouty, obstinate and subject to migraine, mistaken about organic molecules, but a great scientist nevertheless.

7. France in Turbulent Times: cartoons showing the events of 27–29 July 1830 leading to Louis-Philippe's accession: 'Bravery rewarded'. In the first, the pharmacist says, 'Close the shop. Here come the rebels', and in the second, 'In conscience, I could not let them win'.

8. Jean-Baptiste Dumas, caricature by Daumier. A patrician. The caption tells us that, since Dumas has become a minister, he has always taken care to avoid the debates, his pretext being that he is too occupied in analysing the speeches of the others.

9. Auguste Laurent, in the outfit of a poverty-stricken hermit, which he was.

10. Charles Gerhardt. 'They called [him] a brigand, and he certainly looked like one' (*Odling*).

11. Closing page of a letter from Laurent to Gerhardt dated 30 May 1847. According to Gerhardt's son, the drawings by Laurent are of Balard ('Who was discovered by bromine'), Gerhardt as Queen Victoria, Pelouze, Dumas (including a two-faced mask satirising his hypocrisy), Laurent himself (probably) and Thenard, who once violently threw Gerhardt out of his house.

12. Hofmann and his students at the Royal College of Chemistry, *c.* 1855. 'Gentlemen, new bodies are *floating* in the air.'

13. Stanislao Cannizzaro. Soldier, politician and family man who still found time for chemistry.

14. Edward Frankland. A distinguished career blighted by the stigma of illegitimacy.

15. Archibald Scott Couper. Suffered a mental breakdown and disappeared from the scene so completely that, decades later, no one could even remember what nationality he had been. His paper of 1858 lay in Wurtz's desk for months as Kekulé's was published.

16. Alexander Butlerov. Bee-keeper, spiritualist and maker of egg soap, and the man who first used the term 'chemical structure'.

17. Hermann Kolbe. Thought chemical structure theory a swindle perpetrated by anti-German elements, including the French-speaking Kekulé.

18. August Kekulé. Began by dozing on top of a London bus, ended showered with honours.

19. The Ouroboros, 'The dragon that devours its own tail'. A single snake consuming its own tail symbolised the Mercury of the Wise, in which everything is found: 'From the One to the One by the One'. In this example of two serpents devouring each other, the winged snake represents the universal world-spirit, which is sucked from the dew and with which we prepare our salt. The lower snake signifies the virgin earth found beneath the roots of plants, the 'Philosopher's Turf' that the alchemist Armand Barbault dug on the night of the full moon, following precisely the instructions of Pseudo-Eleazar.

20. Germania presenting the structure of benzene to the workers: relief panel on the Kekulé memorial, Bonn. The memorial consists of a large standing statue of Kekulé flanked by two sphinxes.

21. Josef Loschmidt. Author of 'The Masterpiece of the Century in Organic Chemistry'. Too provincial to become famous, perhaps.

have different colours) forms four different oxides, MnO, Mn_2O_3, MnO_2 and Mn_2O_7 (a green oil that gives a purple vapour). In these four compounds its valency is 2, 3, 4 and 7 respectively; in other compounds, its valency can also be 1, 5 or 6. (The valency of oxygen is 2.) The manganese atom is changing its valency in different compounds in ways that were completely mysterious to the nineteenth-century chemist, and have to do with the way the electrons in the manganese atom move from one orbital to another.

Now look at the formulae of some typical simple organic compounds: methane, CH_4, ethane, C_2H_6, ethylene, C_2H_4, acetylene, C_2H_2. Is the valency of carbon 4, 3, 2 and 1 in these four compounds? No, it is not. In organic compounds, the valency of carbon is fixed at 4, and the valencies of the other common elements hydrogen, oxygen and nitrogen are fixed also. The valency of carbon does not vary in different organic compounds. A new explanation, or set of explanations, was necessary. The carbon atoms form chains and rings with each other, without any change in the valency of each individual atom:

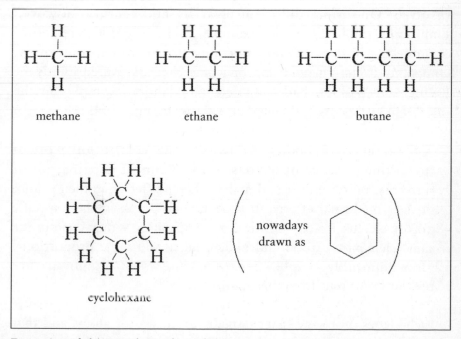

Formation of chains and rings by carbon.

Carbon atoms can also bond to each other using two or three of their valencies:

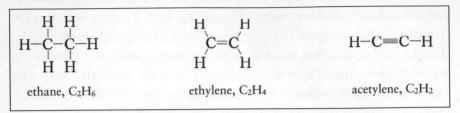

ethane, C_2H_6 ethylene, C_2H_4 acetylene, C_2H_2

Singly, doubly and triply bonded compounds, showing how the valency of carbon remains unchanged at 4.

This accounts for the *apparent* valency for carbon of 2 in ethylene, C_2H_4, and 1 in acetylene, C_2H_2. But the real valency of the individual carbon atoms remains 4. To distinguish these special properties of carbon from the superficially similar but fundamentally different behaviour of manganese and other inorganic elements was a vital necessity.

These concepts were waiting in the wings, and could not emerge onto centre stage until the atomic-weight problem had been resolved. One thing did seem clear by mid-century, however. Empirical formulae alone (for example, C_2H_4, even if it were known accurately, which it was not) were insufficient to explain organic compounds. The phenomenon of isomerism, discovered jointly by Liebig and Wöhler, had spawned many more examples in the intervening years. As Liebig had already written in 1844:

A great class of bodies, known as the volatile oils; oil of turpentine, essence of lemons, oil of balm of copaiba, oil of rosemary, oil of juniper and many others, differing in their odour, in their medicinal effects, in their boiling-point, in their specific gravity etc, are exactly identical in composition – they contain the same elements, carbon and hydrogen, in the same proportions. How admirably simple does the chemistry of organic nature present itself to us from this point of view![9]

Liebig goes on to say that matter is composed of atoms and that the atoms lie side by side and do not interpenetrate each other, and that the properties of the compound depend entirely on this arrangement. But the only kind of different organization of the atoms that he can suggest in the 'compound atom' (or molecule) is one of multiplication; some compounds might contain one of each kind of atom, another two, another four and so on. Even in the context of 1844, this appears lame. Liebig's list of different oil constituents contains so many examples of isomerism that this must be an insufficient explanation; and surely he did not believe that his own fulminic and cyanic acids of twenty years before had formulae that were multiples of each other?

Two main theories about chemical combination had now evolved, which became known as the *radical theory* and the *type theory* already referred to. The radical theory was the descendant of Berzelius's ideas about electrochemical dualism. Arguing by analogy with inorganic compounds, Berzelius believed that all organic compounds could be regarded as made up of an electropositive part and an electronegative part. Thus, for example, in ethyl iodide, the iodine was directly analogous to the iodine in a salt such as sodium iodide, while the radical, ethyl, played the part of the electropositive sodium. By 1850 the theory was under severe pressure from the discovery of new compounds that demanded more and more bending of the rules, but some researches by Bunsen in Germany had appeared to lend it support. Frankland was an adherent to this theory.

In 1845 Frankland, born in Lancashire, moved to London, where he took up an appointment at the Putney College for Civil Engineers and made the acquaintance of a number of English and German chemists. One of these was the formidable and outspoken Hermann Kolbe, a former pupil of Wöhler and the eldest of fifteen children of a Lutheran pastor. Frankland was not a blind believer in the atomic theory, and in this he was probably influenced by Kolbe, who later became vehemently opposed to it.

If the radical theory was correct, the radicals themselves should be just as capable of isolation as the elements. If Davy could obtain potassium from potassium hydroxide, then someone ought to be able to obtain ethyl from ethyl iodide. Frankland and Kolbe travelled to Marburg for a visit to Bunsen's laboratory. On their return they launched into experiments to try to isolate these naked radicals that were predicted by the radical theory. The preparation of a radical had been a goal of Liebig, Dumas and others for at least ten years, without success. Frankland and Kolbe decided to use metals to try to remove the electronegative part of the compound.

July 1848 was hot and sunny, and on the 28th of the month Frankland, working alone on the roof of the college in Putney, sealed some ethyl iodide in a tube with zinc metal and exposed it to heat and sunlight. A reaction took place, producing white crystals and a liquid, but he was temporarily unable to analyse the product, because some of his apparatus had been destroyed in an explosion, and Bunsen had just invited him for another stay in Marburg. Undeterred, Frankland packed up his tube and took it all the way to Germany with him.

This was an action that would certainly have got him arrested as a terrorist today. The tube now contained the world's first sample of a highly reactive organo-zinc compound, though Frankland did not know it. In Marburg, he added water to a tube containing a little of the new compound he had made, to see the effect. A greenish-blue flame several feet long shot out of the tube, leaving an abominable smell in the air. Bunsen, who worked on compounds of arsenic, rushed in and, thinking that the smell was of one of his compounds, told Frankland that he was already irretrievably poisoned and would die. Much to their relief, a test on the black residue left everywhere by the flame confirmed that it was of zinc, not arsenic.[10]

Frankland opened a tube of his compound more carefully under water to investigate the products formed. To his delight, the violent reaction liberated a huge volume of gas. Analysis showed that the main component had the elemental composition corresponding to the expected ethyl, C^4H^5. The other product was the expected zinc iodide, and Frankland announced the isolation of the first radical.

The discovery sparked an intense debate; there was no doubting Frankland's experimental results, but what exactly was the gas that he had succeeded in preparing? It took a number of years of further research throughout Europe to resolve this problem. The debate was bedevilled by the familiar question of the atomic weight of carbon. If it were 12, not 6, the formula of the gas was C^2H^5 or a multiple thereof, not C^4H^5. This was the correct answer. As Laurent and Gerhardt predicted, the gas was not ethyl at all, but a hydrocarbon, butane, C_4H_{10}.

Frankland's experiment was highly significant, but not for the reason that he at first thought it was. What he had tried to do was to prepare 'ethyl', as follows:

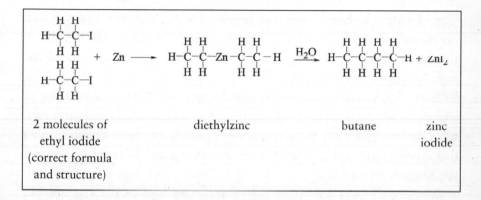

$$\begin{cases} C^4H^5 \\ I \end{cases} + Zn \longrightarrow C^4H^5 + ZnI$$

ethyl iodide   ethyl   zinc iodide
(incorrect formula,
Berzelius dualistic
representation)

Whereas what he had actually done was this:

2 molecules of   diethylzinc   butane   zinc
ethyl iodide   iodide
(correct formula
and structure)

Despite Frankland's misinterpretation of his results, this, together with his later syntheses involving tin as well as zinc, was a key experiment. Not least, he had discovered the first compounds containing an organic group joined to a metal, and had founded organometallic chemistry, which was to become increasingly important with the passage of time. But, in addition, it threw the whole debate about molecules wide open. Despite, or perhaps because of, Frankland's adherence to the basically incorrect radical theory, his work gave a focus for a productive exchange of ideas. The true value came, not in considering the nature of the gaseous product, but in accounting for the nature of his organozinc compounds. In setting out to vindicate Berzelius's radical theory, he had come up with a series of compounds that could not be properly explained by it. How could his 'Zincethylium' (diethylzinc) $(C^4H^5)Zn$ be a neutral substance made of the electropositive ethyl radical and the even more electropositive metal?

Four years after his first experiment, Frankland wrote: 'no matter what the character of the uniting atoms may be, the combining power of the attracting element, if I may be allowed the term, is always satisfied by the same number of these atoms.'[11] This was the first clear statement of the idea of valency. In saying this, Frankland realized that he was, despite himself, moving away from the radical theory, and edging closer to the alternative type theory that was currently running on parallel tracks to it. Frankland says that he feels the two theories might be capable of converging, although the way in which they eventually converged was by the effective victory of the other theory.

The alternative theory that had emerged was the type theory based on the work of Gerhardt, the adherents to which became known in England, rather pleasingly, as 'typists'. The Rubicon between the typists and the radicalists was not an international boundary. Williamson, Odling and other English chemists were typists, as were Kekulé and the Paris-based Wurtz. It was his fellow-countrymen who most neglected Frankland's work; some of them probably just did not get on with him. Chemists overseas, especially Wurtz, recognized his value.

We have already seen how the type theory had its early genesis in work by Dumas, Laurent and Gerhardt. By 'types', Dumas meant substances like his acetic acid and trichloroacetic acid, where the fundamental nature of the substance remains the same, even where one element replaces another, in this case chlorine replacing hydrogen.

Gerhardt, with his comprehensive reassessment of all organic compounds, recognized four fundamental types: the ammonia type, the water type, the hydrochloric acid type and the hydrogen type.

The ammonia type he based on the results of an extensive series of research carried out by Hofmann on benzene and its derivatives soon after his arrival in London in 1849. These included aniline (benzeneamine), the highly valuable intermediate now available on a large scale from benzene as a result of the research by Hofmann and his students. The water type resulted from work by Williamson, who recognized that alcohols, such as ordinary ethyl alcohol, and ethers, such as Gessner's ether, can be formally regarded as derivatives of water. The idea, dating from Lavoisier and Berzelius, that the attempted synthesis of organic compounds was a futile exercise had by now been overturned, and the work of both Hofmann in establishing the ammonia type and of Williamson in establishing the water type was based on synthesis rather than degradation. In doing this, chemists had stepped outside the boundaries of descriptive science and begun to synthesize substances unknown in nature – a profound development.

These relationships between the 'types' can best be shown by giving the modern structural formulae for amines (ammonia type), alcohols (water type) and ethers (also water type). Gerhard's 'hydrochloric acid type' is represented by the organochlorine compounds which we have already considered, in which a chlorine atom replaces a hydrogen in the hydrocarbon skeleton.

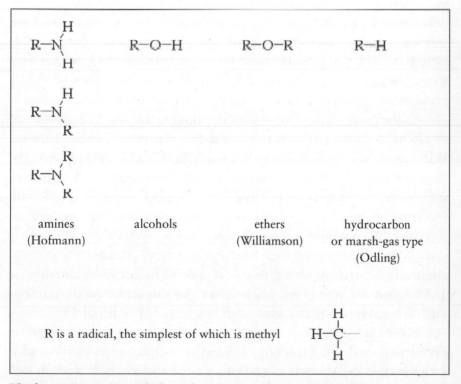

The four main types on which modern structure theory was founded.

Another, and even more significant, type, was first recognized by Odling: the methane, or 'marsh-gas' type, a clearer development of Gerhardt's 'hydrogen type'. Methane, CH_4, is the simplest organic compound of all. It is the smallest member of the vast series of hydrocarbons, compounds containing only carbon and hydrogen, from which the millions of organic compounds now known are constructed. The carbon–hydrogen bond is stable and electrically neutral, because the electronegativities, or electron-attracting powers, of carbon and hydrogen are almost the same. A hydrocarbon molecule is the neutral, stable 'scaffolding' to which other groups, called functional groups, can be attached: –O–H in the case of alcohols, –Cl in the case of Dumas's organochlorine compounds, and so on. Methane, as its alternative name marsh-gas might suggest, is common and had been long known; today's piped household gas from under the sea consists of little else. The formula CH_4 was first correctly assigned to it by Gerhardt. Odling was the first to appreciate that this unreactive and seemingly rather boring gas and its close relatives were of fundamental significance.

It seems such a short step from recognizing these types to drawing the structural formulae shown on the previous page. But, for the time being (we are talking about the years around 1853, which was when Gerhardt made his classification, three years before he died), none of the protagonists was equipped to do so. Gerhardt himself considered his types intellectual constructs only, capable of explaining chemical reactions but having no physical reality. Williamson was a firm believer in the atomic theory but strongly objected to the use of valency bonds in formulae. Frankland was wedded to the radical theory and was sceptical about atoms. All of them were handicapped by uncertainties over the molecular formulae. Who would resolve the impasse?

Archibald Scott Couper's career gave a brief flash of illumination in 1858 before he was gone, his intellect the subject of an irreparable mental breakdown that stayed with him until his death in 1892.

Couper had not taken up the study of chemistry until the age of about twenty-four, following a Scottish education in philosophy and the classics. He was a frequent visitor to Germany and picked up an interest in the subject while there. In 1856 he went to Paris to study with Wurtz, like Gerhardt an Alsatian with a German-sounding name.

Couper only ever published a handful of papers, all in the course of hardly more than a year. It was the third paper on which his claim to fame rests. This paper, entitled 'On a New Chemical Theory', had a most unfortunate history.[12] It was read to the Académie des Sciences at its session of June 1858, but Couper had submitted it to Wurtz several months earlier and it had sat in Wurtz's drawer for the intervening period.[13] It was only through the intervention of Dumas that it made the June session. Couper's paper contained such revolutionary ideas that Wurtz needed time to think about it, but he also thought Couper brash and unreliable.

In his paper, Couper agrees with Gerhardt that the atomic weight of carbon should be doubled, although for obscure reasons he sticks to 8 for oxygen. He criticizes Gerhardt rather vigorously for advancing his type theory, which, although empirically true, advances no explanations; here it is Couper the philosophy student speaking. After giving a clear description of valency using the term 'affinity of degree', he goes on to make the following key statements;

> I propose at present to consider the single element carbon. This body is found to have two highly distinguishing characteristics:
>   1. It combines with equal numbers of equivalents of hydrogen, chlorine, oxygen, sulphur, etc. [In other words, its valency is constant.]
>   2. *It enters into chemical combination with itself.*
> These two properties in my opinion explain all that is characteristic of organic chemistry. The second property is, so far as I am aware, *here signalized for the first time.* What is the link which binds together 4, 6, 8, 10, etc. equivalents of carbon and as many of hydrogen or oxygen . . . *it is the carbon that is united to carbon.*[14]

Couper thus clearly defines the phenomenon of catenation, the power of carbon to form stable bonds with itself, resulting in the chains and rings from which organic molecules are made, and which have been drawn a billionfold by scientists all over the world ever since.

He then gives drawings of the formulae, rapidly becoming what we would now call structures, of organic compounds. He starts with the type that he calls nCM⁴ (he does not define this symbolism) and that we would call fully saturated compounds, that is those not containing multiple bonds between any of the atoms. He uses dashed lines to show the linkages – as in this representation of 'ethylic alcohol' (ethyl alcohol or ethanol).

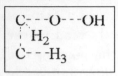

Couper's representation of ethyl alcohol from his paper of 1858.

To convert these to modern structural formulae we have to perform only one major operation, and that is to adjust the atomic weight of oxygen, converting Couper's two oxygen atoms to one (see (a) below). The rest is mere cosmetics and results in the structural formula for ethanol as drawn by Couper's colleague and fellow-Scot Alexander Crum Brown around 1864 (shades of Dalton's balls here) (see (b)); this was later universally simplified by the omission of the circles to give the representation of ethanol familiar to every chemistry student (see (c)).

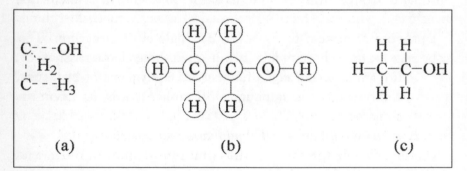

(a)   (b)   (c)

Representations of ethyl alcohol according to (a) Couper, (b) Crum Brown and (c) modern chemists.

Couper never completed a promised fourth paper, in which he proposed to treat what he called nCM⁴–mM² compounds. He would probably have recognized these as what we would now call unsaturated compounds, those containing multiple bonds between one or more atoms.

Couper complained bitterly to Wurtz about the unreasonable delay in the submission of his key paper. As a result, Wurtz apparently told him to pack his bags, and he abruptly left for

Scotland, where he succumbed to mental instability. Because of his abrupt disappearance from the scene, he became virtually a forgotten figure until posthumously resurrected by Anschütz in 1908. It is said that, when Anschütz began to research Couper, he was unable to find anybody who could even say what nationality he had been.

Anschütz also became the major biographer of the man whose paper on the same subject as Couper's became much more celebrated.[15] Tragically for Couper, on 19 May 1858, during the period that his paper was lying in Wurtz's desk, he was stunned to see another entitled (in German) 'On the Constitution and Metamorphosis of Chemical Compounds and on the Chemical Nature of Carbon' appearing in Liebig's *Annalen*. The author of this paper was August Kekulé, possibly the most controversial chemist of them all.

An unnecessary diversion that occupied the minds of many chemists in Britain at around this time, just as the structure theory was being born, was the re-emergence of anti-atomism in the hands of Sir Benjamin Collins Brodie.

Brodie, born in 1817, was the wealthy son of the physician of the same name, who was Queen Victoria's doctor and among other things gave evidence at the trial of William Palmer for poisoning by strychnine in 1857. Brodie the younger studied with Liebig at Giessen, then returned to London, where he built a private laboratory, although his research was rather moderate. After he had become Professor of Chemistry at Oxford, in 1865, he spent nearly all his time on theoretical speculations and abandoned experimental work.

Dalton's atomic theory had always failed to appeal to those who wanted to make chemistry mathematical. Instead of the precise laws that his contemporaries demanded, Dalton had produced 'causal explanations and naive models'.[16] The opinion that all science could with sufficient work be reduced to mathematics is still often encountered among mathematicians and physicists today. Most chemists and biologists beg to differ. One can no more predict the existence of quinine and the plant that produces it from mathematical axioms than the presence of tins of paint in one's local DIY warehouse can predict the existence of David Hockney. One can perhaps explain quinine after the event using mathematics, but that is not quite the same thing.

In the 1860s, Brodie attempted to show by means of a 'chemical calculus' that all chemical substances and reactions can be deduced from mathematical first principles without the need for atoms.[17] He considered atoms to be an unnecessary obstacle between observation and explanation; we must 'free the science of chemistry from the trammels imposed upon it by accumulated hypotheses'. We cannot ask what water is, only what it does, or what it becomes. Science cannot explain nature to us; it can only describe. In this he, as well as Williamson and others, had fallen under the influence of the French positivist philosophers Condillac and Comte.

The theory had success in 'predicting' things that were already known, but the attempt was essentially sterile, of no use whatsoever in taking science forward. Brodie was wasting the time of a lot of people in Britain. Worse than that, he had diverted the intellectual energies of some of the best academic people into a blind alley. It was perhaps no coincidence that the locus of the next great developments in chemical theory would be back on the Continent. Brodie published the first part of his 'Calculus of Chemical Operations', summarizing his theory, in 1866; it received 'a respectful but animated reception [followed by] notoriety and final oblivion'.[18] *Chemical News*, in one of the outstandingly misguided book reviews of all time, devoted a whole issue to it and called it 'The Chemistry of the Future', which must have encouraged Brodie to press on with Part II of his calculus, to appear in 1877. But even by 1867 many of the events to be related in the next chapters had already taken place.

Prince Albert had died in 1861; Hofmann went back to Germany in 1865. Chemistry in the island state returned for a while to its former state of somnolence.

TWELVE

The Kingdom of the Two Italians

A thousand facts in themselves do not alter the progressive state of science, while one fact that has become comprehensible ultimately outweighs them all.

Justus von Liebig, *The Development of Science among Nations* (1867)

The man who had held the key to organic chemistry in his pocket for many years did not live long enough, quite, to witness the full impact of the transformations that his insight would produce. Amadeo Avogadro died in 1856 after a full life; he was nearly eighty. The mystery is why the importance of the work of this highly talented man was for so long neglected.

Lorenzo Romano Amadeo Carlo Avogadro di Quaregua e di Cerreto came of a prestigious family of Turin lawyers, and for many years studied the law himself, becoming a doctor of law in 1796. From 1800 he took up mathematics and physics.[1] Not a great deal is known about his personal life; he was an ugly man who nonetheless apparently had a reputation as something of a womanizer. As might, perhaps, be expected of someone who was an ecclesiastical lawyer by background and training, he seems to have been secretive; little known in Italy, hardly the sort of person who would have been a habitué of international conferences, or a prolific correspondent like Liebig. This must have influenced the amount of time it took for the implications of his excellent scientific work to sink in, but we cannot really blame Avogadro. It was the chemists' fault. For years they could not see what was under their noses. This was also partly because Avogadro, being what we would now call a physicist, published papers entirely full of theoretical material, making it easy for the chemists with their love of experiments to overlook them.

The hypothesis that Avogadro published in 1811 held the solution to the atomic-weight problem. Published in the *Journal de physique*, not a publication likely to be read by the majority of chemists, it was also written in French that was clumsy and often obscure. But, although it has often been stated that Avogadro based his hypothesis only on the physics of gases and not on their chemistry, this is not true. The opening sentence could not be clearer: 'M. Gay-Lussac has shown in an interesting memoir that gases always unite in a very simple proportion by volume, and that when the result of that union is a gas, its volume also is very simply related to those of its components.'

What Avogadro showed quite clearly is that, *if you assume that the smallest particles of an element may be made up of more than one atom*, all the apparent anomalies in the gas laws disappear. Gay-Lussac's law applies to both the starting materials and the products, and any necessity for 'half-atoms' disappears. The molecule of an element may be made up of more than one atom.

This explains perfectly, for example, the reaction between oxygen and nitrogen to give one volume of nitric oxide, which had so puzzled Dalton and the others. The atoms merely swap partners, and the volume remains the same, as shown on p. 96.

Avogadro's terminology is not what we would recognize today, because the term 'molecule' had not been properly invented in its modern sense. He uses the term 'elementary molecule' to mean atom, 'constituent molecule' to mean a molecule of an element, composed of one or more atoms of the same kind combined together, and 'integral molecule' to mean the molecule of a compound, made up of two or more atoms of different kinds. But, allowing for this, his explanation was clear: 'The number of integral molecules in any gas is always the same for equal volumes, or always proportional to the volumes.'

So molecules, not atoms, of a gas always occupy the same volume (at the same temperature and pressure). Gay-Lussac's law is correct, not only about the volumes of reactants, but also about the volumes of the products. It is just that you have to think in terms of molecules, not atoms. Equal volumes of gases contain equal number of molecules. Dumas was also right when he thought that you could compare the weights of the particles of gases by weighing equal volumes; it is just that, once again, he should have been thinking in terms of molecules, H_2, O_2, CH_4 and so on, and not atoms. The relative weights of the gases reflect the molecular weights, not the atomic weights. It was tragic for Dumas that he led the French chemists in giving up on atoms when they were so close to the truth.

Avogadro did not give a precise explanation of these facts, and one would not come until much later in the century. One possible explanation that he could have come up with is that molecules are hard spheres that pack together, and that they are all of the same size at the same temperature. A strange consequence of this idea would be that the molecule of a compound, say H_2O, would have to be the same size as that of H_2 or O_2. The only way this could happen would be if the atoms did indeed interpenetrate when they combined; the 'sperm-and-egg' model or something like it. Where would this leave Dalton's little hooks? The molecule would presumably have no structure, or, if it did, it would have to be inside the sphere.

In fact, the ideas that Dalton toyed with were more sophisticated than this, and once again this extraordinary man came close to the true answer. He pictured each atom of the gas being surrounded with an envelope of heat or 'caloric' and thus behaving as a body 'one or two thousand times' larger than the atom itself. He pointed out that, whatever the shape of the atom itself, the shape of this 'globule' would be spherical, and the globules would pack themselves into layers like balls of lead shot in a barrel.[2]

This is almost bang on. The atoms and molecules are indeed small, but they are in constant motion. It is the collisions of the tiny molecules that create an effective space around them and account for gas pressure. As the gas is heated, its atoms or molecules move faster and the pressure increases. By creating this collision space around them, the molecules of different gases behave in the gas phase as if they were all occupying the same volume. The actual molecules themselves are all of different shapes and sizes, but very small.

Avogadro's paper was not completely neglected from the time it was written until it was rediscovered, with a blinding flash of insight, in 1860, as is sometimes implied. It was always there in the background, though some were unaware of it and others, like Laurent, attributed Avogadro's ideas to Ampère, who published something similar but not so clear shortly afterwards. The problem in the years after Avogadro was not a lack of curiosity, but rather of a cacophony of ideas. Every chemist had his own ideas about the atomic theory. Many of these ideas were crack-brained and were not based on a true understanding of the issues involved. We have to have some sympathy with the senior generation like Liebig and Dumas, who tired of the din and insisted that all such ideas were, if not doomed to perpetual failure, at the very least premature.

The roots of this 'baleful confusion' went right back to Dalton.[3] He unrelentingly rejected the 'French theory' of Gay-Lussac, and Avogadro's explanation of it. There is some evidence that he came round to it not long before his death, but by then he was no longer an active scientist. He had become a museum figure, wheeled out from time to time to give his standard lecture on his unproven atomic theory, but his views no longer carried any weight. More crucially, Berzelius knew of Avogadro's hypothesis but misinterpreted it. He failed to pick up his crucial definitions and ascribed to atoms what should have been ascribed to molecules. It so happens that all the common gases Avogadro used as his examples are diatomic: H_2, O_2 and so on. Berzelius incorrectly interpreted this to mean that he was saying *all* gases are diatomic, which fails to resolve the dilemma.

The view that it was cacophony and not silence that was the problem is witnessed by the fate of a paper by Marc Gaudin that was sent to the Académie des Sciences in 1831. Gaudin was another somewhat peripheral figure, not part of the Paris establishment, and not really a mainstream chemist; he was more interested in photography. Much later he wrote an eccentric book, in which he tried to relate the structures of molecules to the structures of crystals, with bizarre results.[4]

One could not ask for a nicer set of diagrams than those that Gaudin sent in with his paper. They resemble those already shown on pp. 95–6. Using Dalton's atomic symbols, he showed how the volumes of both the reactants and the products can be reconciled.

And yet his paper too sank almost without trace. Berzelius reviewed it unfavourably, and continued to think that no one method of determining atomic weights could give consistent answers, rejecting Avogadro and Gaudin alike. As Partington drily remarks, 'Of the two alternatives, that his formulae were incompatible with Avogadro's hypothesis and hence incorrect, or vice versa, Berzelius naturally chose the second.'[5] Blinded by his electrical ideas, Berzelius could not accept that two like atoms could bond together. Like electrical charges repel.

Organic chemists, without really realizing it, had been led to a series of molecular formulae which were out of step with those used in inorganic chemistry, by a factor of two. The accepted atomic weight of carbon crucially remained as 6 until Gerhardt questioned it. Kekulé for a long time oscillated between the two sets of values. The situation of silicon was even more dire; Thomson assumed silicon chloride to be SiCl (valency of Silicon = 1) and gave an atomic weight of 7.12; Gmelin preferred $SiCl_2$ and 14.25; Berzelius $SiCl_3$ and 21.37; while Odling (*correctly*) thought the balance of probabilities was for four-valent silicon, $SiCl_4$ and 28.5.

Experiment piled on experiment for another fifty years after Avogadro's paper. More experimental results were always useful, if they could be relied upon. But many of them could not, and it was impossible always to know which were reliable. Obtaining accurate analytical results was very difficult. There were many whose results could not be relied on. Chemists moved in a vicious circle.[6] Experimental error in analyses created errors in the formulae. Incorrect formulae led to inadequate theories. Those who tried to make sense of the mess had to award something like Michelin stars for analytical reliability. If an analysis came from the Giessen school, it was probably, but not necessarily, accurate, and so on. This need to criticize others' results led to vitriolic and unproductive arguments in letters and journals, with Liebig usually somewhere near the centre. If two chemists did not get on, the first charge to be levelled was almost invariably that the other fellow's analytical results were misleading.

Laurent reviewed the state of play in 1844:

In considering the rapid progress of organic chemistry in the past few years . . . chemists have reason to be proud of their work; but in glancing towards the future, should they not be alarmed at the researches that are still necessary before the science can be set out with clarity and method? So much time will be required to unscramble the true relationships. . . . New facts and new compounds are published with such rapidity that one can hardly keep up with it because there is no method by which one can classify. One glimpses an era in which each chemist follows his own thread, without concern as to what his neighbour is doing.[7]

The situation was still unresolved in 1859 when Odling wrote:

The majority of English chemists represent the atomic weight of carbon by 6, that of oxygen by 8, and that of sulphur by 16. Dr Frankland would double the atomic weight of carbon, but would retain the old atomic weights of oxygen and sulphur. Mr Griffin, who lays claim to priority in doubling the atomic weights of carbon and oxygen, ridicules the notion of doubling that of sulphur. Dr Williams, Mr Brodie and myself have for a long time [*correctly*] advocated the doubling of all three.[8]

The bracketed formulae based on Berzelius's electrical ideas had by now sprouted a mass of typography – parentheses, buckles, commas, juxtapositions and full stops all struggling to express what were thought to be different levels of affinity and organization even in simple substances like acetic acid. In the early 1860s, Kekulé was able to collect no fewer than twenty-one current ways of writing the formula of acetic acid.

The man who came closest to resolving the situation in these decades of confusion was Gerhardt, when he carried out his thorough review of organic reactions. His strongest piece of evidence came when he looked at all the transformations in which 'carbonic acid gas' (carbon dioxide) took part. He realized that the amount of carbon dioxide released or absorbed in a reaction always corresponded to C^2O^4 and never to half that value. Why did the carbon atoms always go around in pairs? The answer to Gerhardt was clear; they did not. The true explanation was that the correct formula for carbon dioxide was CO_2 and not C^2O^4 and the atomic weight of carbon was 12 not 6. By rectifying this, Gerhardt was actually bringing organic formulae back into line with Berzelius's atomic weights of 1826, although neither of them cared to point this out. But this was still not a *proof*; it was merely the most probable explanation, given on the assurance of a man who had based it on a thorough review of all the literature. Others were free to disagree, and many did. To adopt all Gerhardt's adjustments would be to destroy all the relationships that had been painstakingly built up since the early work of Berzelius, and to annoy Liebig, Wöhler and many others. Gerhardt was accused of doing algebra, not chemistry.

In 1860 August Kekulé organized the first international chemical conference. The main object was to resolve the problems of uncertainty in atomic weights and formulae.

The location chosen was the rapidly industrializing little town of Karlsruhe. The decision to hold it there was based on its central location and the fact that Kekulé's friend Carl Weltzien was professor at the polytechnic and willing to take on most of the organization, but nowhere could have been more symbolic of the new Germany. The town had been founded only since the beginning of the eighteenth century, by the Margrave Karl-Wilhelm of Baden-Durlach. From being a typical provincial German backwater, like Darmstadt, it had rapidly industrialized following the opening of its first railway station in 1843. The 1848 February revolution in France inflamed the new proletariat throughout the state of Baden, who declared the first German republic. This was put down by Prussian troops under the command of Prince Wilhelm, later to become Kaiser Wilhelm I. His equestrian statue is still to be seen in the town, but nearby set into the grass are stone memorial tablets to the twenty-seven citizens of the state who were summarily executed in 1848. In 1998, on the 150th anniversary of the revolution, local students from the Markgrafen-Gymnasium erected a tribute to them in the form of twenty-seven plaster heads. Gerhardt had studied at the polytechnic when it first opened in the early 1830s, and it is tempting to imagine him leaving behind a legacy of discontent when he left.

The conference appears more important today than many chemists judged it to be at the time. Although no fewer than forty-five eminent names were appended to the invitations, this was probably more a tribute to Kekulé's political and organizing skills than to deep-felt enthusiasm on their part. For, of the forty-five, only twenty actually travelled to Karlsruhe, and some of those are today relative unknowns. Bunsen, Wurtz, Dumas and Odling went, but Balard, Brodie, Frankland, Hofmann, Liebig, Mitscherlich, Pasteur, Regnault, Williamson and Wöhler were among those who did not.

However, among the 140 or so participants were several of the rising stars of a new generation of chemists, including the young Dimitri Mendeleev, later inventor of the periodic table of the elements, then twenty-six years old.[9] The periodic table was based on the regular organization of the elements in order of their atomic weights, and so it too could not come into being until the issues to be tackled at the conference had been resolved.

But the man who was to be most remembered as a result of the conference was none of these. He was a little-known Italian from the remote fringes of Europe, the Sicilian Stanislao Cannizzaro.

Had Cannizzaro been born fifty years earlier, he would have entered life a subject of the Spanish Inquisition, for it was not abolished in Sicily until 1782. The island in which he spent his early years remained a lawless and impoverished place sharing its history more with the Arab lands to the south than with the rest of Italy. In the eleventh and twelfth centuries its Norman rulers had kept harems and had run the island with the help of Islamic officers; later, when Sicily came under Spanish rule, the island was run by the Catholic Church. The Church authorities were responsible for punishing crimes, of which murder by vendetta was considered by no means the most serious.

Italy had always been easier to define as a geographical entity than as a political one, and now it was in a state of upheaval, like most of the rest of Europe. Once upon a time the peninsula had been the rich conduit of trade for all sorts of spices, drugs and other goods from the Middle East and the Red Sea. As the ocean routes between Europe and the Americas and the Far East opened up, the centre of gravity of the Continent shifted westwards. Italy had declined in economic power and came under foreign domination. Through lack of coal, it then missed out on the Industrial Revolution, remaining mostly a rural, peasant society. The fate of the various duchies and kingdoms was determined by horse trading between the powerful nation states further north. The Sicily that gave Cannizzaro birth in 1826 was now part of the Bourbon Kingdom of Naples, the largest in area of these many political units and also known, for obscure reasons, as the Kingdom of the Two Sicilies.

Like Avogadro, Cannizzaro came from a fairly powerful family, for his father was a magistrate and minister of police, and his mother was descended from Sicilian nobles. As such, he was one of the 2 per cent of the population of Sicily who could read. Even so, education was under such tight control by the Church that it was a virtual miracle for him to be able to follow such a potentially revolutionary occupation as scientist, and even this was possible only by travelling north to Pisa.[10]

Cannizzaro was politically engaged; no thinking person in the Italy of his youth could fail to be so. In view of the turmoil during the years in which he was trying to build his career, and what happened to him, the wonder is that he was able to be so successful a chemist while at the same time such an active nationalist. The narrative of what happened to him in 1847–9, when he and his contemporaries were still students, is extraordinary.

Italy had been in intermittent revolt against its foreign rulership ever since the end of the Napoleonic Wars, the period known as the Risorgimento. The revolutionary Giuseppe Mazzini founded the movement Giovane Italia (Young Italy) while in exile and returned to spearhead a new revolt that broke out in 1847. Cannizzaro dropped his chemical studies in Pisa and returned to Sicily. Immediately, and at the age of twenty-one, he found himself commanding an artillery battery at the siege of Messina, a desperate affair in which a Dr Palasciano was imprisoned, and nearly executed, for treating enemy wounded. When Messina fell, Cannizzaro was sent to Taormina to try to stop the advance of Neapolitan troops under General Filangeri. But, after the defeat of March 1849 at Novara and the abdication of Charles Albert, the Sicilians retreated towards Palermo, Cannizzaro's unit fighting a rearguard action. The revolutionary Sicilian government fell and Cannizzaro escaped from Sicily on the frigate *Independente*, which managed to avoid the Neapolitan ships and make its way to Marseilles. From here he made his way to Paris, where he was admitted to Chevreuil's laboratory and published a paper on cyanates.[11]

At the end of 1851 he was able to return to Italy as a professor, first at Alessandria and then at Genoa. Although mostly occupied with teaching, he found time to work on benzoyl compounds and discovered the important Cannizzaro reaction. These results were published in a new journal founded by his friend Piria at Pisa, the first issue of which had the distinction of nearly being banned by the censor, the Cardinal Archbishop, because it contained the potentially subversive sciences of chemistry and physics. It was during this fruitful period that Cannizzaro formulated his ideas on molecules, based on his explanations to his students, and culminating in a published paper of 1858. A short time before he had married Harriet Withers, the daughter of a Berkshire clergyman.

In the spring of 1860, revolution broke out in Italy once more, this time to be crowned with eventual success. Southern Italy rose up in response to events further north, and, although the new revolution was at first crushed by Neapolitan troops, Garibaldi with his famous thousand landed in Sicily on 11 May and succeeded in capturing Palermo. At this point Cannizzaro, now thirty-three, dropped everything once more and joined the second expeditionary force under General Medici, although this time he did not see any action in the campaign. In Palermo he became a member of the Extraordinary Council of the newly proclaimed State of Sicily.

There must have been two Cannizzaros. How on earth could he have found the time, or the mental equilibrium, to arrive at a chemical conference in Karlsruhe on 3 September of the same year? The man was a prodigy. How could he have sat through a three-day meeting with a lot of northern Europeans, complete with a steering committee, to which he was elected, as well, no doubt, as leg-pulling and academic jokes, with all that must have been going on in his head? The united kingdom of Italy had not even been declared yet.

The conference began by appointing its steering committee, which included Kekulé, Wurtz and Cannizzaro.[12] The committee drew up a list of questions that were ripe for discussion, and on the second day there was a lengthy session. The chair was taken by Dumas, now a senior figure although hardly senile at sixty years of age. But in a discipline that had developed so fast during his career, he was a member of a conservative old guard. Cannizzaro later said: 'Habits of the spirit are often the most powerful obstacle to the progress of science. Dumas, accustomed to praise all the works of Lavoisier, for which he had a patriotic worship, considered the dualistic system one of the linch-pins of modern chemistry. He thought acceptance of Gerhardt's system an offence to the memory of Lavoisier.'[13] This may have been a trifle hard on Dumas, given that the type theory had begun with his own research, but by 1860 his memories of

Gerhardt must have been at the very least ambiguous. In 1869 he was to give a lecture on the recent history of chemistry in which he mentioned neither Laurent nor Gerhardt.

On this second day, 4 September, the meeting tried strenuously to clarify what was meant by the key terms *atom*, *molecule*, *equivalent* and *radical*, but without success. The meeting was scheduled to run for only three days, and time was running out. When the final day began, the steering committee posed three questions. One was symptomatic of the atmosphere of failure. Should chemistry return to the principles of Berzelius? Cannizzaro argued strongly against. Gerhardt had shown the right path.

In 1858, Cannizzaro, the devoted disciple of his fellow-countryman Avogadro, had published his essay in Italian, *An Outline of Chemical Philosophy*, which contained this clear statement of what is meant by the term atomic weight: 'The atomic weight of an element is the least weight of it present in the molecular weight of any of its compounds.'[14]

Compounds of hydrogen are known, Cannizzaro said, where the molecule contains one half-molecule of hydrogen, others where it contained two, others three. But no compounds were known that contained less than half a molecule. Therefore, echoing Lavoisier, the half-molecule of hydrogen was entitled to be considered as the atom until proved otherwise. Owing its intellectual genesis to Avogadro and Gerhardt, this statement encapsulated what Cannizzaro had evolved over the past several years in teaching his students. Gerhardt had nearly arrived there with his pronouncements on carbon dioxide, but he had not quite carried the day. Cannizzaro threw things on their head. The attempts of Dalton and, especially, Berzelius to derive atomic weights from first principles and thus construct the molecule, were reversed. The path to atomic weights was through the molecule, not the other way round.[15] The 'chemical molecule' measured by many equivalent weight experiments, and the 'physical molecule' revealed by experiments on gases, were one and the same thing, as Avogadro had stated.

The conference, though, broke up inconclusively. During the discussion that followed Cannizzaro's presentation, one of the delegates, Strecker, announced his intention of using the new proposed atomic weights. In a move that must have had Dumas grinding his teeth and Berzelius spinning in his grave, he said that the original resolution had carried the name Gerhardt, not Berzelius, and that the steering committee had incorrectly changed it. Kekulé agreed, but typically with reservations. Others argued that scientific questions should not be decided by a vote. Before the meeting broke up, though, Angelo Pavesi, professor at Pavia and a Cannizzaro supporter, gave out copies of Cannizzaro's paper of 1858 describing his theories.

Two of the delegates at least were greatly impressed by Cannizzaro's contribution. One of them, Lothar Meyer, read his pamphlet again and again: 'I was astounded at the light thrown on the most important points at issue. It was as though the scales fell from my eyes; doubt vanished and was replaced by a feeling of peaceful certainty.' Mendeleev too was immensely persuaded by the proceedings and the performance of Cannizzaro, which seemed to show him how one person with a clear vision could make a profound impression.

It would be a mistake to assume that the scales fell from everyone's eyes in so dramatic a fashion, or indeed at all. Many of the copies of Pavesi's pamphlet must have been consigned to the waste bin, or filed away forgotten. Lothar Meyer's widely read book based on Cannizzaro's principles was not published until 1864, and even then there were conservatives. Another five years after that, Mendeleev was bemoaning: 'Is it so long since many refused to accept the generalizations involved in the law of Avogadro and Ampere?'[16]

After the conference, Cannizzaro became Professor of Chemistry at Palermo and then at Rome. His later chemical research was moderate, but he had very little time for it. He had become a liberal senator of the new Italy, helping to shape its constitution and framing laws on public health, the education of women and all the other desiderata of a modern state. His teaching was still his life and he continued it until the age of eighty-three, just before he died in 1910; his body was carried in procession from the lecture hall to the cemetery by his students.[17]

Kekulé, disappointed by the inconclusive outcome of the conference, travelled back to Ghent, where he was professor, and thought. And perhaps dreamed.

THIRTEEN

The Man on the Clapham Omnibus

Very little advance in culture could be made even by the greatest man of genius if he were dependent for what knowledge he might acquire merely on his own personal observations. Indeed it might be said that exceptional mental ability involves a power to absorb the ideas of others, and even that the most original people are those who are able to borrow the most freely.

W. Libby (1917), cited in Mellor (1922)

For whosoever hath, to him shall be given, and he shall have more abundance: but whosoever hath not, from him shall be taken away even that he hath.

Matthew 13: 12

On the evening of 13 June 1847, a widow and her son and daughter living in Darmstadt noticed flames flickering out of the large house opposite, the residence of the aristocratic Goerlitz family. The flickering eventually died down and the family thought no more of the incident. It later emerged that this was a fire lit to destroy the body of the Gräfin Goerlitz, who had been battered to death.

Investigation of the case dragged on as a result of an incompetent police investigation; at first, the Gräfin's husband was accused; then it was thought to be a case of spontaneous combustion, with Liebig called as an expert witness to pour scorn on this idea. Eventually, three years later, the servant Stauff confessed to having murdered the Gräfin to get his hands on her property. The family jewellery, which he had disposed of, was traced, and included a family heirloom ring made of gold and platinum snakes intertwined. The ouroboros, as it was known, was an alchemical symbol in use from the sixteenth century and before.[1]

Fifty years after Stauff's confession, the widow's son from the house opposite rose to his feet at a Berlin banquet to respond to the speeches made in his honour, and to describe how inspiration about the structure of benzene had come to him in the form of a dream of intertwining snakes. The family opposite the murder had been the family Kekulé, and the son looking out of the window at the flames was August.

The liquid in the tube brought from England to the banquet and held up to general view by William Armstrong, the President of the Chemical Society of London, was Faraday's original sample of benzene. The name for benzene in German is *Benzol*, and the object of the *Benzolfest*, as it later became known, was to celebrate the effective foundation of a vast new industry as a result of Kekulé's insights into chemical structure and bonding in the 1860s. Intertwining snakes, according to Kekulé's account, held the secret of benzene, and benzene held the secret of Germany's success.

Bringing together hundreds of men (and their wives, if only in an ante-room) in such pomp, the 1890 feast was something above the normal run of conference dinner. To those present, it was a pleasant occasion, but it was also the chance to parade national pride and revel in their membership of the scientific establishment. In 1890 the German Chemical Society, representing the practitioners of the most important science in Bismarck's burgeoning Second Reich, was showing its muscle. It was telling the world that the embodiment of the new technology was standing before them as he rose to speak last. When he did so, Kekulé himself described the meaning of the gathering in somewhat more modest tones: 'The ambitious little body of chemists, proud of its past and full of hope for the future, felt the need, in our jubilee-addicted century, to hold a jubilee on its part also.'[2]

In this sentence there is a hint of one of the difficulties in evaluating what Kekulé truly thought and was laying claim to: understatement. German speakers were, and sometimes still are, prone to employ the figure of speech known as litotes. This manner of delivery was in contrast to, and grew up in conscious opposition to, the high-flown hyperbole that the French had always been accused of. (Even in 1763, on the same day that Dr Johnson kicked his stone, he had upbraided Boswell for using exaggerated language, and Boswell had remarked how such a habit was seen everywhere, but that anyone who had travelled in France had been struck with innumerable instances.[3]) In its worst excesses, the countervailing litotes can produce an effect of oleaginous false modesty, or even menace. Hitler was especially fond of it. Kekulé's speech is a tricky mixture of litotes and pride. To some extent it succeeds in putting over his desired effect of modest bonhomie, but at the same time it is a masterpiece of ambiguity. A countervailing taste of Kekulé in vainglorious mode is given by what he said next, in a passage that was omitted when an English-language translation was published: 'Never before as long as science has been practised has a living person been feted in this manner by his colleagues. Never before has an anniversary been celebrated for a scientific work after only twenty-five years.'

The man was clearly an egoist, but this is not a crime. The controversy that has raged ever since about him concerns not whether he was conceited, but whether he stole the achievements of others. Some of these, like Edward Frankland and Alexander Butlerov, are less well known than himself; others, like Archibald Scott Couper and Josef Loschmidt, until recently were largely forgotten. The anti-Kekulé camp say that these scientists are lesser known or forgotten precisely because of what Kekulé and his friends said or did before, during and after the banquet.

In his speech, Kekulé made the first reference to the way in which the concept of the structures of organic molecules had come to him in dreams a quarter of a century before. But contemporary newspaper reports of the speech make only slight reference to any remarks Kekulé made about dreams, and it seems that he greatly elaborated his account for subsequent written publication.

There are several possible explanations for the dreams. One of course is that they occurred more or less exactly as Kekulé related them, or rather wrote about them. A second possibility is that they were a pleasant fantasy invented by Kekulé to entertain the guests at the banquet, and have no particular significance. This is certainly the impression we get from his initial reaction, when he was asked to submit a written version. He was irritated that a stenographer had not been present, and that he was being asked to write down the 'stupidities' that he had spoken off the cuff.[4] It seems that he then changed his mind and used the written account of his speech to clarify what he had wanted to say and reinforce his claim to a place in scientific history.

But clarifying or manipulating? Did Kekulé use the occasion as a golden opportunity to deny others a share of the credit, and were his 'dreams' an invention to enable him to do so? To claim that an insight comes to you straight from your subconscious is an effective way of saying, first, that it *is* an insight, and not a painstaking piecing-together of bits of evidence, and, secondly, that it is *yours alone*.

Friedrich August Kekulé was born on 7 September 1829 in Darmstadt, Liebig's birthplace a generation before.[5] The name Kekulé sounds vaguely French, but the family Kekule were originally Protestant Bohemian nobles. In 1620 one of his ancestors took part in a bloody anti-Hapsburg rebellion and as a result went into exile in Germany. Nearly two centuries later, August's father seems to have added the acute accent to prevent the family name's mispronunciation by French speakers. This later backfired, for, with the rise of intense nationalism later in the nineteenth century, the French-sounding name became a handicap. There has even been painstaking (but not very convincing) research aimed at showing that his son August progressively disguised or dropped the accent as German nationalism increased and it became important for him not to be seen as a suspect French-influenced internationalist.[6]

At first August studied architecture at Giessen, but there he fell under the spell of Liebig and took up chemistry. Later he was to claim that his architectural studies had influenced the way that he thought in pictures, just as Laurent's early crystallography had done.

A feature of Kekulé's early career was his mobility. He went to study at Paris with Dumas and Wurtz. There he became a particularly close friend of Gerhardt and attended meetings organized by him in 1851–2. (By this time Laurent was gravely ill.) By the end of the 1850s he had taken advantage of the greater ease of travel round Europe to work with, or visit, most chemists of note, a precursor of the jet-setting academic of today. 'Originally a student of Liebig, I had become a student of Dumas, Gerhardt and Williamson. No longer did I belong to any one school.'

At first, Kekulé does not appear to have been a particularly able student, or at least not in Liebig's view. His initial researches with the master in 1847–50 were pedestrian, and, when he wrote in 1853 from an unexciting post in Switzerland, Liebig was unable to take him back. Instead, he got him a stay in London with a former pupil of his, Stenhouse, at Bart's. According to Kekulé's later account, he could see from Stenhouse's publications that he was a second-rater, or what was graphically known as a *Schmierchemiker* ('messy chemist'),[7] but nevertheless Kekulé decided to spend eighteen months in London to learn the language and to visit his stepbrother. But the implication is that Liebig found Kekulé uninspiring, and Kekulé's justification was written forty years later.

Whilst in London, he met the British chemists who were struggling with the problem of the nature of organic compounds – Williamson, Odling and others. It was here on a summer evening in 1854 that Kekulé claimed to have had the first of his dreams.

During my stay in London I resided for a considerable time in Clapham Road in the neighbourhood of Clapham Common. I frequently, however, spent my evenings with my friend Hugo Müller in Islington at the opposite end of the metropolis. We talked of many things but most often of our beloved chemistry. One fine summer evening I was returning by the last bus, 'outside' as usual, through the deserted streets of the city, which are at other times so full of life. I fell into a reverie, and lo, the atoms were gambolling before my eyes! Whenever, hitherto, these diminutive beings had appeared to me, they had always been in motion; but up to that time I had never been able to discern the nature of their motion. Now, however, I saw how, frequently, two smaller atoms united to form a pair; how a larger one embraced the two smaller ones; how still larger ones kept hold of three or even four of the smaller; while the whole kept whirling in a giddy dance. I saw how the larger ones formed a chain, dragging the smaller ones after them but only at the ends of the chain. I saw what our past master Kopp, my highly honoured teacher and friend, has depicted with such charm in his 'Molekular-welt', but I saw it long before him. The cry of the conductor 'Clapham Road' awakened me from my dreaming; but I spent a part of the night in

putting on paper at least sketches of these dream forms. This was
the origin of the 'Structural Theory'.

The reference to Kopp's whimsical book about molecules
gambolling about is unnecessary. Despite the light-hearted tone,
Kekulé is a man obsessed with priorities.

Simultaneous discovery is, of course, by no means rare in science.
It is probably the norm. An earlier example was the discovery of
oxygen. Priority for this seems to go to Scheele. The notebook
describing it is undated, but appears to be from somewhere in the
period 1770–3. His manuscript describing his results was not sent to
the printer until December 1775 and not published until 1777, when
it appeared with a preface by his friend Bergman claiming priority
for him. In the meantime, Joseph Priestley had discovered the gas
independently on 1 August 1774, and Lavoisier made it soon
afterwards. Lavoisier certainly learned of oxygen by reading of
Priestley's experiments, but neither of them knew about Scheele.
Some have claimed that Priestley must have heard of Scheele's results
from Bergman with whom he was in correspondence, but there is no
evidence for this. Priestley categorically denied it.

Another episode occurred only a few years later, when Henry
Cavendish, James Watt (better known for his steam engines, but he
knew his chemistry) and Lavoisier have all been credited with
establishing the nature of water. The crucial experiments were
Cavendish's, but the phlogiston idea prevented him from clearly
describing what he had found. Watt was the first to state clearly that
water was a compound of two components, and Lavoisier, using his
new theoretical framework, went on to state that it is a compound of
hydrogen and oxygen. In this case there seems no doubt that Lavoisier
unethically tried to claim priority. He and other members of the
French Academy were astonished when they first heard of the others'
results, and Lavoisier repeated Cavendish's experiment only because
the other members of the Academy told him he ought to check it.

Perhaps we can see the events of the 1860s as a replay of those of
the 1770s, with Kekulé cast in the role of Lavoisier. There is no
doubt that Kekulé was a highly accomplished chemist; his other
researches provide ample proof of this. But, like Lavoisier, he was
skilled at presenting facts in a way that emphasized his

contributions. For example, here in 1867 is Kekulé minimizing the significance of his predecessors, burying them all in an avalanche of faint praise: 'Scarcely any chemist at the present day speaks of Radicals or of Types; and yet the type-theory, like its forerunner the dualistic view [the radical theory], had its good points. It would, I think, be wise not wholly to lose sight of these several theories, which, after all, are based on a considerable number of facts.'[8]

Establishing the truth about Kekulé's claims to priority for the structure theory has generated vast heat over nearly a century and a half. Russell reviewed the various claims.[9] There were four concepts that had to be recognized.

The first concept was that *each element (in organic compounds) has a fixed valency.* Russell gives at least equal credit to Frankland alongside Kekulé, and there seems little doubt that Kekulé failed to acknowledge Frankland properly. But it is not clear that Kekulé did this deliberately. Because he was a follower of the type theory, he seems to have had a psychological difficulty in communicating with those who held to the essentially incorrect radical theory. The two met at least twice; once in London and once at Wurtz's house in Paris, but extraordinarily, as far as can be made out, they never seem to have discussed chemistry. Frankland, in his autobiography, left unfinished at his death in 1899 and published in bowdlerised form in 1902 as *Sketches from the Life of Edward Frankland*,[10] pays tribute to Kekulé but rejects his claim to priority. Unfortunately, in re-editing his collected works, Frankland made numerous modernizations and other amendments to make them clearer. There is no evidence at all that he used the Lavoisier method of retrospectively grabbing priority, but the unintended effect was to make it difficult to work out what he claimed to have done when.[11] Kekulé presumably thought Frankland an eccentric who had nothing to teach him. But 'a discovery made by an adherent of the radical theory . . . does not become the property of the first adherent of the type theory who happens to translate it into the new atomic weights'.[12]

At the beginning of his *Benzolfest* speech, Kekulé does give credit to Frankland. He begins, echoing Newton and others, by saying that we all stand on the shoulders of our predecessors. Fifty years earlier, he says, the stream of chemical progress had divided itself into two branches; the first, chiefly flowing through French soil, had been the verdant product of Laurent and Dumas. The other, following signposts set up by Berzelius, led for the most part through boulder-strewn territory. But in the end, the two streams had converged, with Frankland at the head of the contingent following the Berzelius (radical) stream.

The second concept was that *polyvalent atoms can couple together two or more parts of a molecule*, such as oxygen in R–O–R (an ether; R = a radical). Russell gives priority (just) to Williamson with his studies on the 'water type',[13] although Williamson did not use the exact phrase 'holding together', while Kekulé did.

The third concept involved acknowledging that *carbon is tetravalent*: the atom has four 'hooks'. This generated a most complex priority dispute. Kekulé became involved in an argument with Kolbe and Frankland about who first recognized this vital fact, while others advanced the claim of Couper, who by then was too ill to argue for himself. The question of date priorities between Kekulé, on the one hand, and Kolbe/Frankland, on the other, is only a matter of weeks; both teams published several papers over a period of a couple of years (February 1857 to June 1859).[14] It comes down to a highly technical dispute about exactly what each laid claim to in the series of papers – complicated by the fact that Kolbe and Frankland used the old atomic weights while Kekulé had for some time been switching about between the two systems. In his first paper of 1854 he used Gerhardt's atomic weights, but by 1857 he had unaccountably slipped back to $C = 6$, $O = 8$.

And yet this heated dispute was largely a sterile one, because the realization that carbon is four valent had in fact already been made two years earlier by Odling, who never made any claim to priority at all! Odling's recognition that there was a 'Marsh-gas' type, based on methane, CH_4, was given in a lecture at the Royal Institution in 1855, but it was never published except in the Institution's journal. Kekulé's biographer asserts that Kekulé never saw the publication.[15] This may be true, but Kekulé was a close associate of Odling around that time, and it seems highly unlikely that he did not pick up the idea from the undemonstrative Odling.

The last of the four key concepts was the recognition that *carbon can bond with itself* to make rings or chains. This phenomenon, called catenation, is perhaps the most important of all in defining organic chemistry, and we have already dealt with its history; credit must be given equally to Kekulé and Couper, although Couper was for many years overlooked and Kekulé did nothing to rectify the injustice. How ethical was it for Kekulé, who on his own claim had 'immersed himself anew in thorough studies of the pioneers', never to mention Couper?

It seems extraordinary that Couper was forgotten by the British chemical community and that it was Anschütz, Kekulé's biographer, who rediscovered him after Kekulé's death. There is, however, a poignant side issue that complicates the picture and does not really reflect badly on Kekulé.

Couper's second main paper contained the practical results of his work on the reactions of phosphorus compounds with salicylic acid, the acid from willow bark. It was largely on this work that he based his bonding theories. Kekulé tried to repeat this work. The fact that he took Couper seriously at first is evidenced by the fact that he tried the reaction no fewer than twenty times but was completely unable to duplicate the results. Couper was now unable to speak up for himself and was forgotten. The implication of Kekulé's experiments was that Couper's work was all unreliable. Then, more than twenty years later, Anschütz tried the reaction again and was able to reproduce it. Kekulé had let the solution stand for too long, and the compound on which Couper had staked his reputation had decomposed. Such is the thin boundary between fame and obscurity.

There is also the thorny problem, now virtually impossible to resolve, of the extent to which each of the claimants to fame thought that his formulae truly represented the molecule in space, the 'little thing', and to what extent he continued to think that they were just convenient representations to explain reactions, a kind of algebra. Gerhardt had explicitly rejected spatial reality. He thought that we can understand only reactions, and that it was meaningless to ascribe a particular structure. Williamson too was a sceptic under the influence of the positivist philosophers, and Frankland did not even completely believe in atoms. The vital insight is that molecules have a three-dimensional structure, and that the properties of a compound are determined by this structure *and nothing else*. When a chemist assembles two carbon, six hydrogen and one oxygen atom in a certain way as represented by the paper formula for ethanol, the product will be ethanol with exactly the same properties as every other sample of ethanol that has ever been made. There is no need for vital force or any other additional factor. The structure alone is all that is required, and there cannot be two ethanols with the same structure. If the properties of a compound are determined by its structure and nothing else, the need for a vital force vanishes like mist on a hot day.

As far as Kekulé is concerned, it seems that he too did not come to realize that his formulae were descriptions of an underlying physical reality until after his most important papers had been published, and it was others who pushed him into saying it. This is where Alexander Mikailovich Butlerov comes in.

In the pursuit of chemistry, Russia, as in so many other ways, remained a backwater until the Crimean War. After that defeat many professionals in Russia woke up to the fact that the country was slipping behind the rest of Europe. But it was an uphill struggle with many obstacles. The French Revolution of 1848 and other turbulence in the West caused the Russian authorities to impose a ban on all foreign travel, which lasted until 1855, and as a result there was little real contact with new ideas. Chemistry was taught over-academically, and until the 1860s there was hardly any practical research done.[16] At this time, a group in St Petersburg, including the composer Borodin, who was also a Professor of Chemistry, became active.

Several hundred miles away to the south-east there was Butlerov, a lonely self-taught figure born to a family of landowners that was probably originally German or English. In this isolated corner of Europe, teaching at the university of Kazan, he dabbled in various scientific subjects and was by no means committed to being a chemist. At one point he went into business, making soap out of egg yolks. The venture failed because he could never get his soap as yellow as that of his commercial rivals. It was only later that he found out that what they were calling 'egg soap' was merely ordinary soap dyed yellow. Academic advancement in Russia was still based mostly on teaching ability and there was no need for him to do original research; facilities at Kazan, like those at Bordeaux or Montpellier, were meagre. It was Butlerov himself who later introduced the 'Liebig model' into Russia, and helped make chemistry the strongest scientific discipline there until 1917.[17]

Butlerov, although highly thought of in Kazan, was unable to get permission to go abroad until 1857–8. This trip, which lasted a year and on which he was accompanied by his wife, seems originally to have been planned as an extended vacation, but, like Liebig and others before him, he found that travel completely changed the direction of his life. He visited Germany, where he met Kekulé and learnt much from him during long conversations. Then he went to Paris for five months, the only place that his schedule allowed much of a stay. Everything else had been a Cook's tour. He needed an exceptional mind to assimilate so much so quickly and even to come within shouting distance of the West European chemists. But he did much more than that. After visiting Paris, he published eighteen papers between 1861 and 1863.[18]

Butlerov, bee-keeper, spiritualist and autodidact, was the first to introduce the term 'chemical structure' (replacing Gerhardt's vaguer 'constitution'). He expressed the concept at a time when there is strong doubt about whether Kekulé had clearly framed the concept even to himself. His formulae, Butlerov said, represented 'la position topographique des atomes'. He also recognized more clearly than anyone else the nature of the double bond, realizing that the linkage in compounds like ethylene, $H_2C=CH_2$, uses up two of the valencies of each carbon atom. This, rather than being a source of increased strength and stability, is the opposite, resulting in greater reactivity as the double bond opens up in reactions. The double bond is a functional group.

There is still plenty of scope for debate about Butlerov's actual contributions. It comes down to trying to decide to what extent he framed the *concept* 'Chemical structure' and to what extent he merely framed the *expression*; and to what extent he *learned* from Kekulé, and to what extent he *taught* him. There have been so many different opinions that Alan Rocke had to divide them into four camps: strongly pro-Kekulé, weakly pro-Kekulé, strongly pro-Butlerov and weakly pro-Butlerov.[19] The strongly pro-Butlerov camp claims that Kekulé immediately changed his terminology after hearing a presentation by Butlerov at a conference in 1861. Rocke himself came down for Kekulé. It is certainly true that Butlerov had virtually no knowledge of, or interest in, the structure question until his travels thrust him into the centre of the debate, and that most of what he learned, he learned from Kekulé. There was also the influence of Couper. At first, in 1859, Butlerov was criticizing Couper's views in print, but two years later he was crediting Couper, not Kekulé, with being the source of his central ideas.

These uncertainties will probably never be completely resolved. During the Soviet era, Butlerov's claims were vigorously advanced by Russian scientists, and equally strongly rejected by Westerners, with the effect that poor Butlerov became a cold-war political football, several decades after his death.[20]

So, on every side, Kekulé's claims to priority for his structure theory of organic chemistry are surrounded by controversy. Before 1890 he never referred to his dreams, except, according to his son, to the family circle, a claim that obviously cannot be admitted as evidence. Yet his earlier publications, such as a paper of 1858,[21] gave full recognition to the ideas of Williamson, Gerhardt, Wurtz and others.

Kekulé when young was generous. Whether through his own efforts or those of others, or a combination, over the next decades he became generally credited in Germany and abroad with the discovery of something momentous: that molecules had definite structures in three-dimensional space, and that the properties of different organic compounds could be explained solely by their structure. By 1890 he felt it necessary to agree with this general opinion and by bringing up his first dream shifted his insight back to 1854, to dispose of many of the claims of the others.

The growth in nineteenth-century organic chemistry that had begun haltingly in the middle of the century and had accelerated with unstoppable momentum in the decades leading up to the *Benzolfest* was mostly founded on the synthesis of aromatic compounds. Benzene and its relatives – naphthalene, toluene, phenol (carbolic acid), cresols – became readily available on a vast scale from coal tar, and chemical manipulation produced a huge range of wealth-generating products. The list of nineteenth- and early twentieth-century industrial and household products shows the scale of this industry: aniline dyes, trinitrotoluene, carbolic soap, creosote, naphthalene mothballs and so on. With the dyestuff industry leading the way, aromatic chemistry, a term coined by Kekulé, would continue to mushroom over the following decades, producing not only many more industrial chemicals, but important drugs like aspirin (1899), Paul Ehrlich's pioneering antisyphilitic drug Salvarsan (1909) and many others. As Kekulé himself said, accurately if not modestly, the benzene theory was the keystone of the quarter century of chemistry leading up to the banquet.

At the centre of this new technology was Faraday's mysterious substance benzene. Faraday himself had correctly determined that it was a hydrocarbon containing equal equivalents of carbon and hydrogen. Its formula was some multiple of CH, although this could not be absolutely certain until the atomic weight of carbon had been fixed at 12 as a result of the Cannizzaro revolution. Benzene was then recognized to be C_6H_6. Unlike other substances with such a low ratio of hydrogen to carbon, it was so stable that it could be converted into many stable derivatives by substitution reactions, without decomposing. For example, treating benzene with a mixture of concentrated nitric and sulphuric acids gives nitrobenzene $(C_6H_5)NO_2$. The nitro group had replaced one of the hydrogens in benzene, but the benzene molecule itself was not degraded under these harsh conditions. Benzene showed remarkable stability, which seemed to be incompatible with the high degree of 'unsaturated' character implied by its C_6H_6 formula, comparable, for example, with that of the highly reactive acetylene, C_2H_2.

Developing Couper's bonding ideas, his Edinburgh colleague Alexander Crum Brown had suggested this structure for benzene in 1864. This representation is completely inadequate. It does not explain the stability or the unreactivity. Something much more sophisticated was needed.

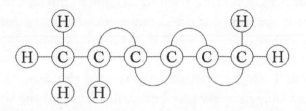

In addition to its stability, the benzene molecule seemed to be highly symmetrical, which Crum Brown's formula was not. There was only one nitrobenzene, one chlorobenzene and so on. If the substitution of hydrogen was continued, using even harsher conditions, there were *three* different dichlorobenzenes ($C_6H_4Cl_2$), later called *ortho-*, *meta-* and *para-*.[22] Continuing the substitution, there are three different trichlorobenzenes, three tetrachlorobenzenes, then only one pentachlorobenzene and one hexachlorobenzene, C_6Cl_6. But at the time, 1865, that Kekulé disclosed his structure for benzene, these facts were only just emerging, and would take several more decades of work to be fully documented. The experimental evidence was not yet all there, and it was a giant leap of the imagination for Kekulé to arrive at the benzene structure on such skimpy evidence. If, indeed, it was him.

Here is Kekulé's account of how the solution to this problem came to him in his second dream.

Something similar happened with the benzene theory. During my stay in Ghent I resided in elegant bachelor quarters in the main thoroughfare. My study, however, faced a narrow side alley and no daylight penetrated it. For the chemist who spends his day in the laboratory this mattered little. I was sitting writing at my textbook but the work did not progress; my thoughts were elsewhere. I turned my chair to the fire and dozed. Again the atoms were gambolling before my eyes. This time the smaller groups kept modestly in the background. My mental eye, rendered more acute by repeated visions of the kind, could now distinguish larger structures of manifold conformation: long rows, sometimes more closely fitted together all twining and twisting in snakelike motion. But look! What was that? One of the snakes had seized hold of its own tail, and the form whirled mockingly before my eyes. As if by a flash of lightning I awoke; and this time also I spent the rest of the night in working out the consequences of the hypothesis.

He quotes Liebig: 'Countless spores of the inner life fill the universe, but only in a few rare beings do they find the soil for their development.' He makes other references to being singled out by providence to receive these unique insights, but it is false modesty. It could in theory have happened to anyone, he is saying, but it happened to happen to me because I am better.

Did, could, this dream (and the first one) have taken part exactly as Kekulé described it? This is one of the most famous unresolved questions in the history of science. Were they dreams, reveries, daydreams or just inventions? Psychologists have been called in to study the account that Kekulé gave. Their opinions are ultimately inconclusive.[23] There seems little doubt that it *could* have been just as he described. There are many examples of insights coming to people in the shady world of the semiconscious state, and modern research on dreams, consciousness and half-consciousness has come up with nothing to make Kekulé's dreams seem implausible. It is the historical record that is unconvincing.

The benzene dream supposedly took place some time in the winter of 1861–2, but the paper fully describing the benzene structure was not read to the Chemical Society of Paris until January 1865. Why the delay? Just like Liebig and Berzelius before him, he says he is against premature speculations: 'We should always leave fruit on the tree until it is ripe.' 'Let us learn to dream, gentlemen, then perhaps we shall find the truth. . . . But let us beware of publishing our dreams until they have been tested by the waking understanding.' So three years of mature reflection were required for him to feel able to publish in black and white something that he said came to him with a blinding flash of inspiration. He says that the completed manuscript lay in his desk for almost a year before work by others led him to publish.

The benzene structure that emerged from the 1860s is a thing of considerable beauty and intellectual satisfaction. Like the DNA structure of nearly a century later, it *had* to be right. Kekulé's symbolism of the snake devouring its tail holds the key.

The carbon atoms in benzene form a six-membered ring in which there are no reactive ends. Two of the valencies of each carbon atom are used in bonding to its neighbours, and a third attaches it to a hydrogen, leaving one valency over to form a second bond with one of its neighbours. According to the nineteenth-century view, the double bonds are fluxional – in a constant state of rapid migration round the ring like a snake chasing its tail, so that at any one instant it is not possible to say whether a particular bond is double or single. This easily explains the symmetry properties of benzene – why there are exactly three dichlorobenzenes, for example. The two potential compounds shown under (c) cannot be distinguished because of this rapid equilibrium.

The unreactivity of benzene is less easily explained by this picture. It is not clear why a compound with rapidly oscillating double bonds should be less reactive than a compound with stationary ones. Benzene continued to challenge the intellect all through the twentieth century. More sophisticated insights eventually arose from quantum

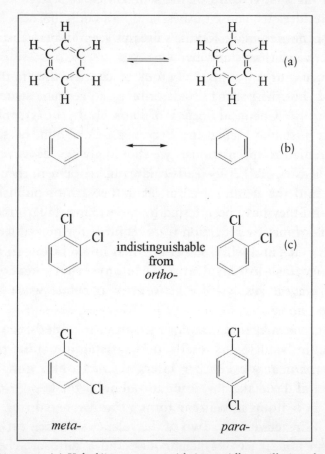

Benzene structures: (a) Kekulé's structure with its rapidly oscillating double bonds (oscillation shown by the symbol ⇌), (b) its modern representation, using the symbol ⟷, which means that the two forms are not in equilibrium but that benzene is a hybrid of the two, and (c) dichlorobenzenes, showing how the hexagon structure accounts for the existence of three, and only three, isomers.

theory, beginning in the 1930s. According to this picture, the bonds in benzene are neither single nor double, but something in between, resulting from overlap of the orbitals of all six adjoining carbon atoms. The electrons in the extra bond are 'smeared out' over the ring. Quantum theory also predicted that, if a series of hydrocarbons was made containing 4, 6, 8, 10, 12 . . . double bonds, then every alternate member (benzene with 3 double bonds, the ten-ring hydrocarbon with 5 double bonds and the fourteen-ring compound with 7 double bonds) would show aromatic stability, while those in between with 4, 8 or 12 double bonds would be unstable and reactive. This was eventually confirmed by making them.

Had the benzene insight occurred exactly as Kekulé described, it would have been one of science's greatest single-handed achievements, and was held out as such well into the twentieth century. It was enthusiastically taken up by Carl Jung, who was very interested in alchemy as evidence of a link to the unconscious mind. He seems to have got a lot of the facts wrong.[24] Then Arthur Koestler called the benzene dream 'probably the most important dream in history since Joseph's seven fat and seven lean cows'.

But, even if we were to accept Kekulé's account at its face value, there are at least two potential barriers on the road to our agreeing that he had thereby established priority for his hexagonal structure. One barrier is sausage shaped, and the other is in the shape of an almost-forgotten Austrian.

Kekulé played with several different types of representations for benzene in the 1860s (see next page). The predominant picture was in fact what has been called his 'sausage' or 'bread-roll' formula shown in (a). In 1865 he first used a hexagon, but without any bonds in it, *and with the hydrogen atoms, not carbon at the corners*. It is not until 1866 that he drew something vaguely approaching a hexagon with three double bonds (b). As Partington humorously remarks, 'Elsewhere only plates of "sausage" formulae appear' until the 1870s.[25] And if we are to award the laurel to the first person to use a hexagon to represent an aromatic compound, however imperfect the realization, then it has to go to Laurent, who had drawn Liebig and Wöhler's benzoyl chloride, rather mystically, as a hexagon back in 1854 (c.).[26]

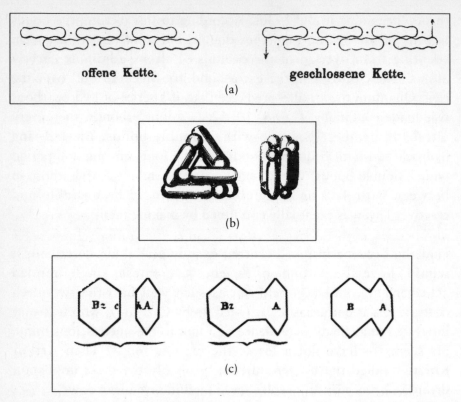

(a) Kekulé's 'Sausage formulae' for open-chain and ring compounds, (b) his 'sausage-formula' for benzene (1866) and (c) use of hexagons by Laurent in 1854 to show the reaction between benzoyl chloride and ammonia.

The unknown Austrian was Josef Loschmidt, who was born in a small village near Karlsbad (Karlovy Vary) in 1821.[27] He studied chemistry and physics at Prague and then Vienna and, after brief spells with unsuccessful industrial concerns, qualified as a schoolteacher in the early 1850s. His career then showed a meteoric rise to professor and senator of the University of Vienna despite the fact that initially (like Liebig) he had no doctorate; he was an excellent physicist and it was principally his physics research that established his reputation. Although apparently a loner, he was by no means an obscure recluse. For most of his life he lived with a much younger woman, Karoline Mayr, and they eventually married when he was sixty-six and a child was on the way. He published only a few papers, mostly in physics, and never went to an international meeting. He died in 1895.

Loschmidt fully understood the importance of Avogadro's work, and the paper for which he is best known deals with just this field of gas physics. Like Avogadro before him but with the benefit of four more decades of scientific progress, he recognized that the 'chemical' molecule discovered by studies of chemical reactions and chemical equivalents, and the 'physical' molecule whose behaviour could be studied in the gas experiments, were the same thing. By using the methods of physics, Loschmidt was able to estimate the number of molecules that occur in a given weight or volume of gas. The standard used was 2 grams of hydrogen (since the hydrogen molecule is H_2), a weight of gas that occupies 22.4 litres at atmospheric pressure. The number of molecules that this weight or volume contains is approximately 6×10^{23}, or 6 followed by twenty-three zeros. This is called either Avogadro's number or Loschmidt's number.[28]

In 1861, four years before the events of 1865 celebrated by the banquet, Loschmidt wrote a little book, *Chemische Studien*, that was brought out privately by a small Vienna publisher. It had only limited sales, although a copy can be seen, for example, in the British Library.[29] Loschmidt's work was never published in a mainstream chemical journal, but it has been pointed out that this begs the question, where could he have published? The best German-language journal available was Liebig's *Annalen*, but the ever-argumentative Liebig had a down on Austrian chemistry, and in 1838 had laid into P.T. Meissner, Loschmidt's teacher, in devastating style in its pages.

The respected information scientist Bill Wiswesser called Loschmidt's booklet 'the masterpiece of the century in organic chemistry'. In it, Loschmidt used what he knew or could infer of the rules governing bonding to draw out structures for over 300 molecules in a mere forty-seven pages.

A comparison between Loschmidt's structures, Kekulé's 'sausage formulae' and modern molecular models is convincing. See, for example, the various representations of aniline shown on the next page. Noe and Bader give many more examples.[30] The key questions are not whether Loschmidt's insights were better than Kekulé's of the same era, which they clearly were, but why Loschmidt was overlooked, and, more importantly, did Kekulé see Loschmidt's structures at the time?

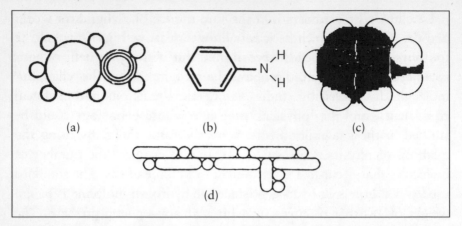

Representations of aniline: (a) from Loschmidt's book of 1861, (b) modern structural formula, (c) modern space-filling molecular model, showing the size and shape of the electron orbitals making up the molecule, and (d) Kekulé's 'sausage-formula' as published in 1866–7. (Noe and Bader 1993)

The first question has much to do with 'the Matthew effect' – the tendency for received opinion to emphasize a small number of perceived geniuses. Once a reputation has achieved a certain critical mass, it tends to be a self-fulfilling phenomenon. For both personal and political reasons, Kekulé's reputation grew unstoppably for many years. He was a distinguished and productive professor at the heart of his profession; Loschmidt was not. Loschmidt was first forgotten, then rediscovered by Anschütz when he wrote Kekulé's biography in the early twentieth century, then forgotten again, only to be resurrected more recently. Meanwhile, most of the textbooks continue to give all the credit to Kekulé.

Anschütz asserted that Kekulé never saw Loschmidt's book, but that he knew about his formulae. Kopp, Kekulé's friend and associate, wrote a review of it in Liebig's annual review of chemistry for 1861. It might seem possible for Kekulé to have seen the review but not bothered to get hold of a copy of the book, but this seems unlikely, given his usual assiduity. And it is not supported by the facts. Anschütz later found a letter from Kekulé to another chemist friend Erlenmeyer where he refers to Loschmidt's 'confusion formulae'.[31]

Kekulé published his hexagonal benzene structure in two important papers in 1865, the year commemorated by the banquet. One was in French and contains a footnote in which he says that he prefers his representation of benzene to those of Loschmidt or Crum Brown, while in the other, essentially a translation into German, the footnote is missing. Some have argued that Kekulé must have been subconsciously influenced by having seen Loschmidt's formulae, but to the unbiased observer, the influence seems much more direct than that. At this point, we feel that we have heard enough of Kekulé's subconscious. Loschmidt joined the list of Kekulé's non-persons and was not referred to again.

Others have proposed that there might be another psychological reason for Kekulé to come up with his account of his dreams. He was well travelled and multilingual. At this time of rapidly increasing German nationalism, he had come under attack for being unpatriotic and under the influence of the suspect French. At the head of the chauvinist pack was the strident Kolbe.

In 1870, just as the structure theory was becoming firmly established, Kolbe fortuitously became editor of the *Journal für practische Chemie*, and for the rest of his career used it as an anti-structure platform, attacking bonds, rings and chains. His editorials also 'began increasingly to carry partisan political, religious, nationalistic and personal overtones'.[32] Kolbe was carrying out a personal vendetta, which accused Kekulé not only of adhering to foreign-influenced 'Jesuitism and Popish hierarchy' (an appalling development of Liebig's gibes against Dumas a generation before), but of being a fantasist and illogical thinker. In a letter to the English chemist Roscoe, he described Kekulé as uneducated and unable to think clearly, and also as lying, stealing and seeking to derogate great men such as Berzelius. He had deliberately falsified history in order to put himself and the French chemists, of whom he was so fond, in the foreground.[33]

A lesser man than Kekulé might have crumpled under such a prolonged and vitriolic attack. Kekulé did not; he had the support of most German chemists of a more liberal frame of mind, and most thought Kolbe mad. But Kekulé had not been psychologically strong enough to stand out entirely against the pressures being put on him, and the increasingly intolerant general climate in Germany. On the outbreak of the Franco-Prussian War in 1870, he had written to a friend, 'So it is war. . . . What a shame! What bastards [*Hundevolk*, literally 'dog-people'] these French.'[34] It must have come as a

Chasing the Molecule

tremendous psychological boost to him later, when his views on benzene, and on bonding generally, had gained universal acceptance. So there is a theory that his 1890 invention of the snakes was a posthumous kick in the teeth for Kolbe, who had died in 1884. Kolbe said that structures were often assigned by chemists 'in their sleep', with no justification for them. 'You thought me a fantasist,' Kekulé is saying. 'But see how my fantasies are applauded.'

There is also the feeling that Kekulé was a gifted man with a more artistic temperament than most of his fellow-chemists, and who did not feel wholly at home in the society of literal-minded and philistine industrialists. Perhaps his dreams were a way of pointing out that he was not quite like them. Reciprocally, there is nothing the philistine likes more than the knowledge that an artist sees his worth. Kekulé was the ideal person to be lionized; this cultured and sensitive man, the chemists were able to say to the world, is one of us. See how our technology not only makes us rich, but makes us dream.

Kekulé spent much of his later years chasing honours. After ten years of effort and the expenditure of considerable amounts of money, he was able to drop the acute accent from the surname when he became ennobled by the Kaiser as Kekule von Stradonitz in 1895, a year before his death. He had become a member of all the great scientific academies except one; only the Czechs held out. Despite an entreaty by Kekulé's son that the family had come of Bohemian stock, they were not likely to admit as a foreign member someone who had by then become a convinced German nationalist.[35]

We will return to the banquet and let Kekulé have the last word.

Someone has spoken of genius and the benzene theory has been called a work of genius. I have often asked myself, 'What is a genius and what a work of genius?' . . . It is said that a genius recognizes the truth without knowing the evidence for it. I do not doubt that even in ancient times this kind of thinking occurred. . . . It is also said that genius thinks in leaps. Gentlemen, the growing intellect does not think in leaps. That is not given to it.

I modestly disclaim being a genius, he is saying. There is no such thing as a genius. But if I am wrong and there is such a thing, then clearly it is me.

FOURTEEN

All is Found!

Let the Looking-Glass creatures, whatever they be
Come and dine with the Red Queen, the White Queen and me!
Lewis Carroll, *Through the Looking-Glass, and*
What Alice Found There (1872)

Kolbe saw the whole of structure theory as a swindle. Worse, it was a swindle being promulgated by German chemists, not the hated French. He noted that the French had leaped ahead in the earlier years of the century by avoiding the influence of the dreaded *Naturphilosophie*, while the scientists of his beloved Fatherland had been seduced by its suspect charms.[1] If the Germans were not careful, their chemistry would be ruined by structure theory and the French would get the upper hand again. He used his editorship of the *Journal für praktische Chemie* as a vehicle for violent polemics against valency theory, saying that he regarded all attempts to determine the positions of atoms in space as 'vergeblich' (futile).

Kekulé was to him particularly suspect because of his cosmopolitan leanings, hardly preferable to the Frenchman Wurtz, who had famously commenced a large chemistry textbook with the words: 'Chemistry is a French science. It was founded by Lavoisier of immortal memory.' (The Germans got their revenge; when a statue to Wurtz was erected in Strasbourg during the period of German ownership, his name was inscribed as 'Würtz'; after 1918, when Alsace became French again, the authorities cut out the 'ü' and replaced it with a 'u'.[2])

Meanwhile, other chemists began to build molecular models. In 1865 Hofmann festooned the lecture theatre of the Royal Institution with models made from croquet balls and sticks.[3]

Rocke has pointed out a wonderful psychological mismatch between Kolbe and the new structure theory.[4] In Berzelius's old radical theory with its brackets and braces, there was a sort of hierarchy between the atoms. The molecule was organized like an army unit, with some atoms having organizational superiority over others of the same element. But, in the new picture, every atom of the same element was of equal status, such as in the benzene ring of six equal citizens. It was democracy on the submicroscopic scale, a bitter pill for the authoritarian, xenophobic and anti-Semitic Kolbe to swallow. Most chemists were by now considering the benzene ring as what Kolbe called 'infallible dogma' as 'the pope is for catholics'. He could not swallow this.

It was a battle fought partly in the subconscious. Although Kolbe fulminated against the fantasists like Kekulé, he had no insight into the psychological imperatives that drove his own opinions. A psychoanalyst might have helped. 'Scientific matters must not be taken personally. I cannot help myself, I *must* criticize and contest the chemical ideas that I consider false and worthless. . . . I ask you, did Liebig ever hesitate to express his convictions?'[5]

Kolbe was comprehensively wrong, but in 1870 there was as yet no absolute proof that molecules had a structure, although most chemists believed it. To see how that proof came about, we have to enter the looking-glass world.

In Italy they tell a pleasant legend, which, unlike most myths in that ancient country, cannot date back further than the nineteenth century. When God created the world, he made Italy first. Later in the day he realized that he was a country short, and, being somewhat pushed for time, he picked up the mould for Italy and threw it into the sea, where it broke in half and made New Zealand.

It does seem a coincidence that there should be two land masses on opposite sides of the world of approximately the same size, each shaped very like a cavalry boot, with the obvious distinction that one now consists of two roughly equal-sized islands with a strait between, while the other is a peninsula grafted onto the underbelly of Europe.

But there is another important difference between them. If we pick up New Zealand and move it into the northern hemisphere, placing it where Italy should be, the difference becomes obvious. New Zealand is right-handed while Italy is left-handed, or vice versa; which is which is a matter of definition. The two shapes show the

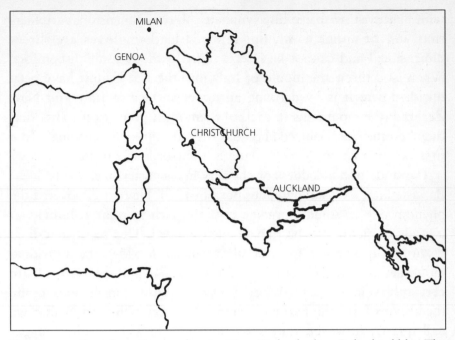

Sketch map of southern Europe showing New Zealand where Italy should be. The two countries are of the opposite handedness, or chirality.

phenomenon of handedness, or *chirality*.[6] They share this characteristic with many of the objects that we see around us; the word *handed* immediately takes us to one of them: our hands, which are mirror images of each other, as are our feet, our shoes and our ears. (In this account I shall use the word 'symmetrical' to define bilateral, or twofold symmetry; there are other forms of symmetry, such as the fivefold symmetry of a starfish or a buttercup.)

This phenomenon of chirality was to play a major role in the eventual realization that the molecule was, indeed, a 'little object', which, despite being too small to be seen, had a character, a shape, a topography, that could be discerned in a logical and satisfying way. It was not a quasi-metaphysical entity. Its structure and inner life could be probed as intimately as a mechanic might probe the interior of a car. In due course the sceptics were put to flight, although Kolbe never accepted the defeat.

Two-dimensional objects cannot be chiral; they are symmetrical and can always be superimposed if necessary by picking one of them up and turning it over. The polythene gloves that are given away free with hair dyes are essentially symmetrical, two-dimensional artefacts that will fit either hand, but tailored leather gloves are three-dimensional and cannot be interchanged without some discomfort. When God threw the mould of Italy into the sea, it must have been upside-down. It is by defining an upper surface to Italy and New Zealand (the mountains in each of them must point to the sky) that they become chiral objects, because they are three-dimensional. To a first approximation, New Zealand is a mirror image of Italy.

The study of handedness or chirality in chemistry began with Jean-Baptiste Biot, a crystallographer born in 1774. He first discovered the phenomenon in some inorganic crystals, such as quartz. Individual crystals of quartz can be right- or left-handed. This seems puzzling, given that quartz is a form of silica (silicon dioxide), the individual molecules of which, consisting of three atoms in a plane, O–Si–O, are certainly symmetrical. When these individual molecules come together to form the particular transparent crystalline form that we call quartz, however, they stack up in a chiral crystalline mega-structure. It has been compared with a builder constructing a spiral staircase out of symmetrical bricks. Individual quartz crystals may be right- or left-handed, purely at random.

Chirality is studied most conveniently by looking at the rotation of the plane of polarized light passing through the sample. Normal light consists of waves that are randomly oriented, but passing them through a polarizing filter allows through only the waves that are vibrating in a particular plane. The concept of polarized light used to be a difficult one to explain, but has been made much easier by the advent of polaroid sunglasses, with which most people are familiar. The molecules of the polaroid plastic function as a kind of tiny venetian blind, letting through only the light waves that are polarized in one plane. Their practical use on the beach is to reduce glare, but we can use them in a simple experiment to demonstrate what Biot found.

If you take two pairs of such sunglasses and look through the lenses of both, orienting one pair so that it is held vertically and the other horizontally, very little light gets through. The planes of polarization of the two lenses are at right angles and all the light is cut out. When the two pairs are aligned with the polaroid filters in the same orientation, the maximum amount of light gets through. Biot used inorganic crystals instead of polaroid plastic, but the effect is the same.

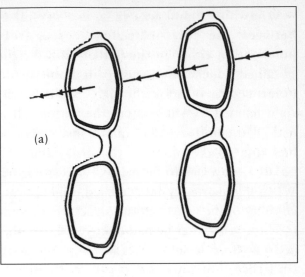

(a)

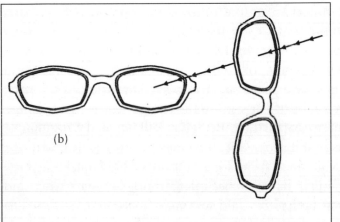

(b)

Two pairs of polaroid sunglasses; (a) with filters parallel, (b) with crossed filters and (c) with a sample of an optically active substance placed between them.

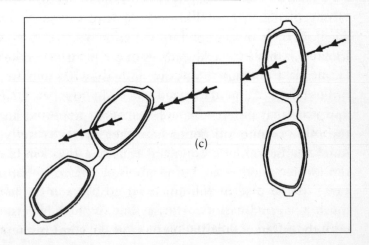

(c)

What Biot found was that, if a crystal of quartz is placed in between the two polarizing lenses, it rotates the plane of polarization either to the left or to the right. (The same result is obtained whichever end of your home-made instrument, called a polarimeter, you look through. Right-handed corkscrews remain right-handed even if you cut the handle off and fix it to the other end.) The chiral crystal of quartz inserted between the two polaroids has rotated the plane of the polarized light so that the nearer polaroid now has to be rotated to restore the maximum light level. When the quartz crystal is melted, this optical activity, caused by the chirality of the quartz crystal structure, disappears.

Three years after he had discovered this phenomenon, Biot looked at some organic substances in his polarimeter. He was surprised to find that, not only did many of them rotate the plane of the polarized light, but that, unlike quartz, *the optical activity was retained when the crystals were melted, or the organic compound was dissolved in a solvent.* In 1815 he found optical activity in turpentine, laurel oil, lemon oil and camphor, and in 1818 he found it in ordinary sugar syrup, cane sugar dissolved in water.

Berzelius visited Biot in Paris in 1818–19 and learned how to use the polarimeter. Ten years earlier, Berzelius had isolated from muscle an acid, lactic acid, and showed it to be chemically identical with the acid Scheele had obtained from sour milk in 1780. French chemists had for the previous thirty years been denigrating Scheele's work and insisting that what he had isolated was merely acetic acid; Berzelius showed they were wrong and took up the cudgels on behalf of his wronged countryman. 'I have thought of letting the French chemists swallow their own puffed-up experiments,' he fulminated. But, on looking at the lactic acid samples in the polarimeter, he found that his lactic acid from muscle was not quite the same as Scheele's from sour milk. His lactic acid rotated the light rays; Scheele's did not. It appeared that different samples of what appeared chemically to be the same compound could be either optically active or optically inactive. Berzelius recognized that the two kinds of lactic acid represented another kind of isomerism, related to but not exactly the same as the case of fulminic and cyanic acids. The phenomenon must, he agreed with Gay-Lussac and Wöhler, have something to do with the different distribution of atoms in some mysterious way.

On 22 May 1848 the man whom a recent opinion poll was to name as the most famous Frenchman of all time submitted a note to the Academy of Sciences describing an experiment as significant for the development of chemistry as Davy's discovery of potassium, or Wöhler's synthesis of urea. In each case, the experimenter *knew* that he had done something momentous, and, although Louis Pasteur did not dance around the laboratory like Davy, his mentor Biot certainly almost did so when Pasteur repeated the experiment for him. 'Tout est trouvé!', Pasteur is said to have exclaimed.[7]

Pasteur is now best known for his research in bacteriology and fermentation, and it is these that made him famous with the general public and an icon of humanitarianism. He and others were responsible for putting to flight Liebig's views on eremacausis. Fermentation and infectious disease are due to attack on the tissues by ill-intentioned organisms. The organisms do not arise by spontaneous generation, or happen to chance along once the tissues have already succumbed to a kind of catalytic breakdown caused by the complexity of their molecules, as Liebig thought.

But, like Faraday before him, Pasteur began his scientific career with some ground-breaking research in organic chemistry. Born in the Jura in 1822, he came from a military family; his father was a veteran sergeant who had fought all over Europe with Napoleon I. Intensely religious and patriotic, Louis had arrived in Paris in 1843 to study with Balard at the École Normale and we have already heard his account of hearing Dumas lecture at the Sorbonne.

Soon after he arrived in Paris, Pasteur also met Laurent, who not long before had begun to ponder the possible importance of the third dimension in chemistry. Laurent showed him a sample of crystalline sodium tungstate that appeared to be homogenous, but that on examination under a microscope could be seen to be a mixture of three different types of crystal.

Some earlier thinkers also had toyed with the question of whether three-dimensional geometry might play a part in chemistry. Plato, on no evidence whatsoever, had associated each of his four elements with one of the regular geometrical figures: the tetrahedron (fire), cube (earth), octahedron (air) and dodecahedron (water). Wollaston, an early supporter of Dalton, thought perceptively in 1808 that the difficulties in the atomic theory would not be resolved until such time as 'we shall be obliged to acquire a geometrical conception of the relative arrangement in all three dimensions of solid extension'.[8] He thought that a compound A–B would be linear (which it would have to be; there is no other possible arrangement), while AB_4 would be tetrahedral, with the four Bs arranged at the corners of a tetrahedron or triangular pyramid. He rather disappointingly then backed off and doubted that the geometrical arrangement of atoms would ever be known for certain.

Laurent was one of a very small group of chemists of his period who felt that chemistry was related to crystallography and that the key to understanding it was geometrical. 'It appears to me at this moment impossible to represent by a linear formula an atomic arrangement in three dimensions,' he wrote.[9] In this he must have been influenced by his early interest in crystallography and minerals. Laurent visualized the molecule of ethane, which he thought was C^8H^{12}, as a four-sided prism with the carbon atoms at the apices and the hydrogen atoms midway along the twelve edges. (Why he did not suggest a cube is not clear; Gerhardt raised this point and Laurent responded, but his argument is difficult to follow.[10]) Such an explicitly graphical representation was anathema to Dumas, who felt that any such attempted physical representation was premature without a great deal more experimental evidence. The crude picture of a prism is completely wrong in detail because the molecular formula of ethane is C_2H_6 not C^8H^{12}. But this was effectively the first use by any chemist of a graphical representation of an organic molecule, composed of atoms, and with a three-dimensional structure that might be the key to its chemistry.

Another acid that showed the same phenomenon of variable optical activity as lactic acid was tartaric acid, isolated by Scheele from wine lees, the crystals that acidic wines deposit on standing. Biot showed that ordinary tartaric acid rotates the plane of polarized light to the right. But, as with lactic acid, a different form of tartaric acid, chemically identical but not rotating the plane of polarization, had also been obtained. This was named Racemic acid, after the Latin word *racemus* for a bunch of grapes.

Biot, Mitscherlich and others intensively studied tartaric and racemic acids, and many of their salts, for several years.[11] But, in his famous experiment, Pasteur noticed something that all the others had missed. The crystals of the salt, sodium ammonium tartrate, that he was studying were chiral. He could see that certain tiny faces on the crystals made them right- or left-handed; he was looking through his magnifying glass at what was effectively a pile of tiny right- and left-handed gloves. The crystals of the salt of ordinary tartaric acid were all right-handed (or, strictly speaking, they were all of the same handedness), whilst those of the same salt of racemic acid were an equal mixture of right- and left-handed. Mitscherlich had in fact written a paper on this very salt, and had claimed that the crystals of sodium ammonium tartrate and sodium ammonium racemate were identical.

Using his magnifying glass and a needle, Pasteur carefully separated the right- and left-handed crystals of the racemic acid salt into two piles. When he dissolved one pile in water and measured its optical rotation, it was indistinguishable from that of the ordinary tartaric salt. The other pile gave an optical rotation that was equal in magnitude, but opposite in sign. He had prepared the hitherto unknown mirror-image tartaric acid twenty years before Lewis Carroll's Alice ventured into the same territory.

Pasteur was extremely fortunate. The formation of mirror-image crystals is quite a rare phenomenon, so he was lucky to stumble upon it. If anyone was likely to do so, however, it was the meticulous Pasteur, who had an infallible eye for noticing things that others had missed. But he was lucky in another way too. It was later found that his salt only crystallizes in mirror-image forms below 27°C. Had he been working during a hot spell, the crystals would not have obliged.

Pasteur clearly recognized the difference between the optical activity of the kind shown by quartz (and also some organic crystals), and that shown by his left- and right-handed tartaric acids. The first kind disappears when the crystals are melted or dissolved, and is due to the right- or left-handed arrangement of non-chiral molecules in a chiral crystal; the spiral staircase model. The second kind is unaffected by melting or dissolving, and could be due only to the inherent handedness or chirality of the *molecules themselves*. Exactly how a molecule could be right- or left-handed, Pasteur could not say. Perhaps they were in the form of right- or left-handed spirals, like even tinier versions of the quartz crystal.

If the priority contest Kekulé versus the rest of the world was a scrappy affair, leading to fighting on the terraces and with extra time still being played, the later international match between Le Bel and van't Hoff was an elegant score-draw, played out on lush grass before an appreciative crowd. It is one of the best examples known of genuine simultaneous discovery in science, and settled once and for all the question of whether molecules have a structure.

Remember that carbon is four-valent; it bonds to four other equal or different groups. Wollaston had been the first to speculate what shape a molecule AB_4 might be. The most symmetrical arrangement of four objects around a central point is not as a square, but as a tetrahedron, or triangular pyramid. Butlerov in 1862 specifically proposed that this might be the way in which the four valencies of the carbon atom might be arranged. Even earlier, when the remarkable Dalton asked his friend Mr Ewart to perforate the little wooden balls so that they could be linked, he had told him to arrange the four holes in one of them as a tetrahedron, although Hofmann, in constructing his 1865 croquet-ball models, had incorrectly arranged them as a square.

In 1874 two young chemists of different nationalities were working side by side in Wurtz's Paris laboratory. They seem to have been on amicable terms, but never discussed the proposal for which they were to be most remembered, and which they published quite independently of each other.

One was the Dutchman Jacobus van't Hoff, then twenty-two and destined to have a successful and influential chemical career. Born in Rotterdam and training as a chemist at the University of Leiden, he spent a short time with Kekulé in Bonn, but found the atmosphere uncongenial, so moved to Paris in 1873. The other was Joseph Achille Le Bel, an Alsatian, less well known in later life and a little older than van't Hoff. He had worked in Balard's laboratory before moving to Wurtz's.

The solution to the mystery of optical activity in organic compounds came to them both simultaneously. Van't Hoff published it in a fourteen-page pamphlet in Dutch dated 5 September 1874, and on 5 November of the same year Le Bel presented a paper in Paris coming to exactly the same conclusions, although his paper is slightly more theoretical and not as graphic as van't Hoff's.

The tetrahedron is highly symmetrical. The molecule of methane, for example, consisting of a carbon atom surrounded by four hydrogens joined to it in a tetrahedral array, has numerous planes of symmetry. If we take this basic molecule CH_4 and progressively substitute other atoms or groups for the hydrogens (remember Laurent?), the symmetry decreases. In the compound in which two of the hydrogens have been replaced by two other different atoms or groups (such as CH_2BrCl), there is still one symmetry plane left, passing through the middle of the two halogen atoms.

But now take this a stage further, to the ultimate situation in which every atom, or group of atoms, attached to the carbon atom is different from every other. A remarkable thing happens. We can now make two versions of the molecule which are mirror images of each other. Lactic acid is an excellent example, as shown here.

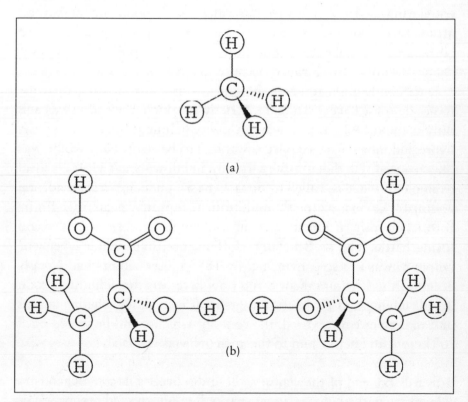

The tetrahedron: (a) the highly symmetrical methane molecule, and (b) right- and left-handed lactic acid. Wedges represent bonds coming out of the plane of the paper, and hatched lines bonds going back into the plane of the paper.

(Pasteur's tartaric acid is a little more complex, so we will stick with lactic acid.) The four different groups attached to the central carbon atom are hydrogen, methyl (CH_3), hydroxyl (OH), and carboxyl (CO_2H), the group that is characteristic of acids. (Remember how Mitscherlich tried to convince Berzelius that all organic acids were made of carbon dioxide joined to a hydrocarbon?) But the principle remains the same whatever the four groups are, provided they are all different. Berzelius's optically active lactic acid from meat consisted of molecules all having the same handedness, while Scheele's from milk was the racemate – a term coined in honour of racemic acid to denote any substance that consists of equal numbers of left- and right-handed molecules, with the optical activity of the two forms cancelling out.

This explanation advanced by Le Bel and van't Hoff not only explained optical activity in organic compounds, but gave convincing evidence that the four valencies or bonds of the carbon atom (each composed of a pair of electrons) pointed towards the corners of a tetrahedron; four different groups pointing around a square do not cause chirality, because the overall molecule is planar.

If its bonds point towards the corners of a tetrahedron, then the molecule must have a structure. Strictly speaking, their idea was still only a hypothesis, but it was an overwhelmingly persuasive one. More and more evidence piled up; it had to be right. Only Kolbe was unconvinced. He thought the virtually unknown van't Hoff an ideal example of what he called a 'metachemist', pursuing 'transcendental chemistry'. But Kolbe's 'malicious stupidity reaching limits fortunately rare, if not unique, in the annals of chemistry'[12] was a losing battle. The Le Bel–van't Hoff picture, like Kekulé's benzene before it, was just so convincing. By 1874 most chemists had already accepted the structure theory, and now here was the definitive proof. To all intents and purposes the great debate about whether atoms and molecules truly existed, the roots of which could be traced back to Dalton, and before him to the ancients, was resolved.

Towards the end of the century, these discoveries in stereochemistry (chemistry in three dimensions) sparked an unexpected reappearance of the vitalism debate in a much more sophisticated guise than it had assumed earlier.

Pasteur was a religious man and a believer in vitalism. He had also pointed out the fact that no reaction carried out by the chemist in his laboratory ever led to an optically active substance. If the product of the reaction was something capable of optical activity, the product was always the racemate. This is because ordinary chemical reagents are incapable of distinguishing left from right, so the product of normal reactions is the optically inactive form consisting of equal quantities of left- and right-handed molecules. To produce just one, or just the other, seemed to require an intelligent intervention. Only living organisms were capable of producing the optically active versions. (Pasteur incorrectly thought at first that biological organisms always produced the optically active form. This is not true; there was already the example of Scheele's racemic lactic acid produced biologically.)

In an address to the chemistry section of the British Association in 1898, the respected scientist F.R. Japp aired these observations of Pasteur made forty or more years earlier, and launched a much more formal debate on their implications.[13] Japp pointed out that the attention of biologists had, naturally enough, focused on Pasteur's later researches on bacteriology, but his earlier, more technical experiments on chiral, or handed, molecules raised matters of enormous importance. These concerned the debate on 'the most fundamental question that physiology can propose to itself' – whether life can be reduced to chemistry alone, or whether a vital force also operates. The concept of vital force had recently fallen into disrepute, he says, because of the crude way in which it had been applied, verging on the occult.

But Japp pointed out that 'living matter is constantly performing a certain geometrical feat which dead matter . . . is incapable – not even conceivably capable – of performing'. This feat is the production of optically active materials. The only possible explanation that he could see was the one involving vital force. Living organisms were able to 'tweak' the synthesis of an organic compound so that only right- or left-handed molecules were produced.

Imagine yourself throwing a large handful of coins into the air over your shoulder. In doing this, you are putting yourself in the position of a chemist who synthesizes a compound with chiral molecules. Roughly half will come down heads and half tails. This is the result that the chemist gets in his laboratory. What would be your reaction if you went away for a few minutes, and when you came back, they were all showing heads? You would conclude that an intelligent agency, perhaps your eight-year-old son, had been at work. Japp's argument was that, although the plants and animals that produced the natural products are not intelligent, yet every day they produce millions of tons of lactic acid, of sugar, of morphine and quinine and strychnine, all of which were optically active and of single handedness. So perhaps there was an intelligent motive force after all.

A considerable debate arose. The well-known physiologist Sir Herbert Spencer was unable to come up with a refutation of Japp's argument, although neither was he prepared to concede the existence of vitalism. In his opinion, both explanations had failed; life is incomprehensible.[14]

Since Japp's time, the debate has moved on considerably, although it cannot be said that it has been completely resolved. The plant cells – yeasts, bacteria and all the other organisms that make chiral compounds – may not be 'intelligent', but this does not mean that they are not capable of churning out molecules of single handedness only. This is because they are already packed full of chiral molecules of their own. Their DNA is shaped like a spiral staircase, and all DNA in the world is of the same handedness. This reflects the fact that its backbone is made up of chiral sugars. This chirality is passed on to all the components of the cell, including the enzymes that are responsible for putting the natural products together. The cells of the plant that makes quinine are incapable of making mirror-image quinine; they are like tiny shoe factories that have only one machine, a machine for making right shoes.

So one set of chiral molecules produces another set, and so on and so on through the whole of creation. But where did the chirality come from in the first place? This takes us back to the beginning of life on earth, and perhaps even before that. In the 1950s, experiments by Stanley Miller and others showed that biological molecules such as DNA and the aminoacids from which all proteins are made could feasibly have been made by heat, lightning and other similar agents acting on the planet's primitive atmosphere. Most of those aminoacids are chiral molecules similar to lactic acid, but containing nitrogen. How could lightning have produced an excess of single handedness?

Two main explanations, or sets of explanations, are possible. One is that the universe itself is chiral, which it does indeed appear to be, but to a very weak degree. According to this hypothesis, the influence on the handedness of the molecules comes from cosmology, and pushes back the explanation to the way in which the universe itself was created.[15] The other explanation is that the chirality of early molecules, either right- or left-handed, arose purely by chance, and that, once one set of molecules had become dominant somewhere on earth, a kind of natural selection ensued that survives to the present day.

Chirality is an important study in modern chemistry. Until the 1960s it was regarded as an academic pursuit, but the situation changed when the toxic effects of the thalidomide drug came to light. Thalidomide is a chiral molecule, like lactic acid, and it was found that, while the desirable drug effects resided in one mirror-image form, the teratogenic properties were associated with the other. So today no drug capable of chirality is marketed without a thorough investigation of both forms. Chemists have become experts at working with chiral molecules.

A modern explanation of how handedness might have arisen on the primeval earth has come from this increasing familiarity with how chiral molecules behave. It is really a sophisticated variant of the randomness explanation, and runs along the following lines. When a chiral molecule is made by conventional chemical methods, it is usually stated that it is optically inactive because the number of left-hand molecules is equal to the number of right-hand ones. This is not strictly accurate. Their numbers are statistically equal, but not arithmetically equal.

Imagine that, in place of your handful of coins, you threw onto the floor the contents of a sack containing a million coins – bearing in the back of your mind that, if you were dealing with molecules, the numbers would not be a million, but Avogadro's number, which is 603,000 million million million. The chance of there being exactly 500,000 heads and 500,000 tails would be vanishingly small. Much more likely would be some irregular combination, such as, for example, 498,993 heads and 501,007 tails. If you now had some mechanism by which head-and-tail pairs were combined and removed (chemically by converting them into something else), you would eventually end up with 2014 tails and no heads.

There are numerous variations on this theme. The debate has been running in gradually more sophisticated form since 1898, and is one of the most difficult questions that science has to address. It currently shows no sign of definite resolution, and gives many hostages of fortune to the creationists.

FIFTEEN

It's All Over for the 'Thing-in-Itself'

If we are able to prove the correctness of our understanding of a natural process by making it ourselves . . . and making it serve our own purposes into the bargain, then it's all over with the Kantian ungraspable 'thing-in-itself'. The chemical substances produced in the bodies of plants and animals remained such 'things-in-themselves' until organic chemistry began to produce them one after another; whereupon the 'thing-in-itself' became a 'thing-for-us'.
Friedrich Engels, quoted in C. Reinhardt and A.S. Travis,
*Heinrich Caro and the Creation of the Modern Chemical
Industry* (2000)

Kekulé's reputation has see-sawed in the century and more since his death, but he shows no sign of becoming a less than controversial figure. In 1930 H.E. Armstrong called for an international 'Court of Appeal' to be set up to deal with the conflicting claims surrounding priority for the benzene structure and bonding ideas generally. This did not take place, but the arguments still rage. A 1990 American Chemical Society symposium to mark the centenary of the famous *Benzolfest* was riven with wrangling and dissent. There were several highly anti-Kekulé contributions and other papers were withdrawn.[1]

The arguments between chemists about the benzene priority are as nothing though compared with the vituperation that resulted when a Freudian psychoanalyst got involved. Alexander Mitscherlich was the great-grandson of the chemist Eilhard Mitscherlich with whom Liebig had had such disputes in the 1830s. In an article in *Die Zeit* in September 1965,[2] he pointed out what he claimed was the obvious sexual content in Kekulé's dreams, especially the benzene dream with its whirling penile serpents. Far from indulging in peaceful reveries in his study facing the alley way (another clear piece of Freudian symbolism), Kekulé had been immersed in fantasies related to unresolved conflicts with his father. (Or was Liebig his father figure?)

This article, even thirty years later, is still the subject of vituperative counter-attack, and deservedly so, especially when we take into account what was actually happening in Kekulé's private life at that time. The Ghent dream, if it occurred at all, cannot be accurately dated, but was probably in 1861 or 1862. In June 1862, Kekulé married Stephanie Drory, a twenty-year-old woman of English descent living in Belgium. After a honeymoon in Switzerland, they visited London, then returned to Belgium, where their son Stephan was born at the house mentioned in the dream, on 1 May 1863. Two days later, Stephanie died. Auguste brought his son up alone. To accuse him of indulging in 'masturbatory fantasies' during this period is a piece of self-indulgence from that most deplorable of decades, the 1960s. Mitscherlich can hardly complain posthumously if his own family life is probed and the conclusion reached that he was transferring the conflict with his own father to Kekulé.[3]

Kekulé's benzene structure, together with the general picture of chemical bonding that he and others developed, has stood the test of time. An alternative possible structure for benzene, the prism structure, was put forward by Ladenburg soon afterwards. In this structure the carbon atoms are at the corners of a Toblerone packet, and all the bonds are single. This neatly explains the existence of three different kinds of disubstituted benzene, but was soon disproved. A structure containing such small carbon rings in such a combination was bound to be unstable. Such was the synthetic ingenuity of organic chemists that developed over the next century that by 1973 it was possible to make Ladenburg's benzene, when it was, as expected, found to be an explosive liquid with none of benzene's abnormal stability.[4]

For the rest of the nineteenth century, the actual nature of the chemical bond that held molecules together remained a mystery, the refinement of which would require advances in therotetical physics that would not emerge until after the First World War. This was the era of 'ball-and-stick' chemistry owing its ancestry to Hofmann's croquet-ball models of the 1860s. The picture that Kekulé and the others had produced, though, was a phenomenally productive one. For the time being it was not necessary to have deep theoretical insights into the nature of molecules to be able to make them in sufficient profusion to change the world.

In 1862 Wöhler, in a letter to Liebig, had already expressed astonishment at the number of chemists being produced by German universities, and wondered what was to become of them all.[5] Three years later, when Kolbe was appointed Professor of Chemistry at the University of Leipzig, he commissioned a laboratory for 132 students. Liebig thought this absurdly large; in 1868 it opened and was immediately found to be too small. Soon afterwards, a writer in an 1875 yearbook observed: 'A veritable army of chemists practises their profession with restless energy . . . where there were scores before, now there are hundreds.' The chemist as a modern professional had arrived. The German word *Beruf* means both 'trade or occupation' and 'profession' and there was less scope for nice social distinction against those who were 'in trade' than there was stereotypically in England or France. Liebig and others had set off a chain reaction.

Over the next century, the science of organic chemistry transformed the world. The average person in the Western world now has a standard of living that 'a mediaeval king could not even have dreamed of', in the words of one organic chemist.[6] There is of course a downside, but there is not enough space in this book even to begin to explore whether, and to what extent, the world is going to hell in a handcart.

Natural products continue to play an important and central role in chemistry. This book has shown how the important new science of organic chemistry mostly grew out of their study. The number of natural products continues to grow year on year; there are now more than 20,000 known alkaloids, and the total number of known natural products is about ten times this figure. The main driving force behind this research, apart from sheer curiosity, is pharmaceutical – the search for new drugs – just as it was in the days of Pelletier and Vauquelin.

In the early days, natural products led the way for chemistry, and showed chemists beginning with Scheele how to improve their techniques. Once the incorrect idea that compounds could not be synthesized was overthrown, organic synthesis developed, but very slowly at first because of the absence of a structure theory (the era of Hofmann at the Royal College, for example). After the structure breakthrough by Kekulé and the others, the pace accelerated phenomenally, but the range of synthetic methods was still rather limited and the plants continued to have the upper hand in what they could do.

During this 'classical' period of organic chemistry, which lasted until about 1950, most new drugs were small synthetic molecules. Working out the chemical structures of an interesting natural product could took decades of work, and its synthesis, even when the structure was known, might be impossible. A good example is strychnine. After its isolation by Pelletier and Caventou in 1818, the correct formula $C_{21}H_{22}N_2O_2$ was determined by Regnault (the figures had later to be corrected to take into account the correct atomic weight of 12 for carbon). But it was then well over 100 years before the structure, the way in which the atoms are joined together, was finally solved in 1948 by a team at Oxford headed by the brilliant eccentric Sir Robert Robinson.[7] Robinson himself had worked on strychnine ever since 1910, when he was only twenty-three.

The number of weapons in the arsenal of the investigators in the classical period was strictly limited. They had to try to degrade the molecule to produce smaller fragments using the mildest possible conditions. One fragment might be identical with a substance already known, either synthetically or by degradation of another natural product. If there was no luck at this stage, the degradation product could be subjected in turn to a variety of other reactions, always with the aim of producing something smaller that could be definitely identified. At each stage it was necessary to argue by analogy from the known reactions of simpler substances to try to explain what was happening, and to suggest what approaches might be fruitful for the next stage of degradation. The criteria for validating this work were extremely limited. If two samples obtained by different pathways had the same melting point and other physical properties, this proved that they were the *same*. It did not prove what they *were*. Everything else had to be proved by logic from first principles. The reactions that turned up during these degradations could not always be understood.

It was often necessary to synthesize whole series of simpler model compounds to clarify what was happening. To add to the difficulty, some natural product molecules could undergo rearrangements of the carbon skeleton even under mild conditions. By 1931, when Robinson reviewed progress, the mere outline summary of what was known about strychnine took twenty-one pages of closely written argument, culminating in a new tentative structure that was not only incorrect, but left three of the carbon atoms still unaccounted for.[8] Robinson's principal rival, Hermann Leuchs, met his death in May 1945 during the fall of Berlin at the age of sixty-five, without ever seeing a solution to the riddle.[9]

The correct structure was put forward as a possibility by Robinson in 1934, but the credit for its definite proof had to be shared with brilliant new arrivals on the scene, the Yugoslav Vladimir Prelog and the American R.B. Woodward. At one point, Robinson clouded the picture still further by publishing another, completely erroneous, structure, in a fit of something resembling panic, which he immediately withdrew and ever afterwards referred to as his 'Frankenstein' structure. So, even after this century of effort, involving hundreds of man-years of effort and some of the world's finest minds, the proposed structure was merely highly probable, not completely confirmed.

All this changed when the modern era of organic chemistry arrived after the Second World War. The new chemistry used physical methods such as various kinds of spectroscopy, to probe the molecule.[10] Within a few years, this enabled the chemist to put his sample in one or two machines sitting in the laboratory, and within a few minutes to obtain an accurate map of which atoms and groups it contained, work that in the classical era would have taken months or years. The chemists of the pre-Kekulé era were like explorers lost in a jungle with measuring rods all of different length. Robinson was like a colonial administrator with his theodolite and tape measure. But by the 1960s there was the equivalent of a satellite positioning system on every benchtop.

The most powerful method of structure determination took another three decades or so to become truly a routine method; this is X-ray crystallography. Since they are hardly bigger than the wavelength of light, molecules are far too small to be seen with a microscope, and can be imaged only using radiation of a much shorter wavelength than visible light. Such radiation became available in 1895 with the discovery of X-rays. In theory it would be possible to determine the structure of the submicroscopic molecule by firing X-rays at a crystal and studying the pattern deflected by the atoms inside. The first successful applications of the method, at first using very simple substances like salt, were made by the father-and-son team of W.H. and W.L. Bragg, for which they won the Nobel Prize in 1915. Forty years of development following the Braggs' experiments meant that by 1956 it was possible to obtain a 'photograph' of the strychnine molecule. This gave not only the chemical structure, the way in which its forty-seven atoms are joined together, confirming Robinson's final proposal, but also the distances between the atoms and the angles between the bonds in three-dimensional space. We now know more about the detailed structure of molecules than could have been dreamed of by the chemists of the classical late-nineteenth-century period. X-ray crystallography is now following ultraviolet spectroscopy, infrared spectroscopy and nuclear magnetic resonance (nmr) spectroscopy out of the specialist research department and into the ordinary research laboratory.

Some organic chemists regretted the advent of these physical methods; it seemed to them like cheating. Robinson himself failed to appreciate their potential power when coupled to computing methods. 'Such computers, costing something of the order of one and a half million pounds, are not likely to become generally available' was his opinion in the 1970s. Organic chemists were proud of their specialized expertise and addicted to the thrill of the chase. The subject gained something of a reputation among other scientists for being an introverted discipline. 'People sometimes ask me the difference between organic chemistry and biochemistry,' said Sir Derek Barton, winner of the Nobel Prize for Chemistry in 1969. 'I tell them that organic chemistry is interesting.'[11]

During the classical era, the great demand was for drugs against bacterial diseases. The majority of new drugs introduced were synthetic, and either were found by completely random testing of new compounds, or were synthesized with only a very hazy idea of how drugs might work. A good example was the sulfa drugs, introduced in the 1930s. These came about because of an idea that it might be possible to kill bacteria with stains like those that the microscopists were using to show up bacteria on their slides. The first successful drug found was indeed an azo dye, but it was then discovered that the azo part was not necessary; it was the non-dye part of the molecule that was effective, and the range of antibacterial sulfa drugs that came into use were in fact colourless. During this era, biochemistry, the complex knowledge of how the cell works, was lagging well behind organic chemistry, the science needed to make the drugs. Medicinal chemistry in those days was like trying to mend a defective television set by throwing different sized and shaped bricks at it.

There was a swing back to natural products with the discovery of the antibiotics; powerful compounds produced by micro-organisms to kill other micro-organisms. Penicillin is the best-known example, but there were many others: streptomycin, chloramphenicol and the tetracyclines, for example. There was no way that the scientists of the 1950s could have drawn out the structure of a molecule and predicted that it would have powerful antibacterial or anticancer properties. The research was from the other direction; investigate the organisms, in this case bacteria and fungi, isolate the bioactive natural products, and then find out what they are. Except that, by the 1960s, finding out the chemical structure had become much easier, and synthetic methods were available to modify the molecule and perhaps make something even better, that the organism itself could not make.

This period after the Second World War was therefore a second golden age for organic chemistry. The new harnessing of physical methods produced another tremendous growth spurt, now largely centred in the new Leviathan, the United States. It was the era of 'Better Living through Chemistry' – Du Pont's slogan, and one that was at that time enthusiastically endorsed by the general public as well as the chemists. Rapacious industrialists and the more aggressive academics marched hand in hand. In his book on the history of DNA, Jim Watson records how his Ph.D. supervisor, the microbiologist Salvador Luria, abhorred most chemists of the era, 'especially the competitive variety out of the jungles of New York City'.[12]

Environmental concerns apart, organic chemistry has now to some extent become a victim of its own success. Some scientists nowadays, echoing Herschel in the 1830s, do not regard it as a true science. There are too few unanswered questions. It has become a technology; a highly successful one, which, in the hands of its best practitioners, can make any desired molecule to order; engineering on a sub-microscopic scale. Within the confines of the giant pharmaceutical companies, these skills are now taken more or less for granted. Impatience with the timescale of traditional chemical synthesis has led to automated mass-production methods producing thousands of new compounds simultaneously for biological testing. They do not even have to be made individually. There are ways of producing controlled mixtures of compounds, a kind of primeval soup that can be tested first, then investigated afterwards if necessary to find out which of its many components has produced the desired effect. Sophisticated computer modelling software then allows the scientist to look at multicoloured images of the way in which the small molecule of the drug interacts with the giant molecules of protein or nucleic acid within the cell to produce its effect. In the best research environments, the chemist is still there as a member of a multidisciplinary team, engaged in activities that would be virtually unrecognizable to a Liebig or a Gerhardt.

At the time of writing, though, this 'combinatorial chemistry' based on the synthesis of many thousands of compounds for biological testing by mass-production methods appears to have yielded rather disappointing results. The problem is that such methods are essentially conservative; they produce large numbers of variants of a known kind of drug, but all with the same basic structure. If you want totally new types of compound that might have new and unexpected drug activities, you need to go back to the chemist.

And, if you need startlingly new kinds of molecules and activities that even the chemist may not have thought of, you need to go back to nature. Natural products are becoming fashionable again as the drug companies try to satisfy their insatiable hunger for new lead compounds – those that show interesting activity even if perhaps they are too toxic for human use. Once you have a promising lead, the skills of the chemist and biologist can be used to try to discover how the drug works, and perhaps modify it chemically to increase its activity and ease of absorption and to reduce its toxicity. Since thalidomide in the 1960s, the path leading to government approval for clinical use of a new drug has become ever steeper and more boulder-strewn.

The way in which scientists from academia and industry track down new natural products has not changed so much since Humboldt and Bonpland's expedition of 200 years ago. Most of the biodiversity in the world is in the tropics, so expeditions still travel to the depths of the Amazonian rainforest to bring back new plants and soil samples for investigation. A shortcut is to enquire of the locals which plants they use against various diseases. Better take along an anthropologist, though, because you will need to know whether the local medicine man is telling you the truth, or is perhaps enhancing his status by telling you what you want to know, or inflating his reputation by pretending to have a greater arsenal of active plants than he really has. And in these days, it might be a good idea to have a corporate lawyer in your party too, because the legal rights of indigenous peoples in the plants and drugs that grow in their territory is nowadays taken very seriously by environmental activists.

About one-third of new drugs introduced, of which only a handful each year get past all the regulatory hurdles, are natural products or result from chemical modification of a natural product. The rest are synthetic. Apart from the rainforest, other rich sources of new drugs have been traditional Chinese herbal medicines, of which many are known, and, increasingly, marine creatures. The sponges, sea urchins and poisonous fish that live beneath the surface of the oceans produce a wide variety of alkaloids and other natural products with which to kill each other and protect themselves from being eaten, and many of them are of startlingly novel structural types. It is true that in some ways the synthetic chemist has the upper hand over nature, because he or she has access to synthetic methods that the plant or animal, which can use only the enzymes that nature has given it, does not. But nature continues to surprise us by showing that an organism somewhere, be it bacterium, crab or orchid, does indeed make something hitherto thought impossible, and perhaps also makes something else of a type that we had not even thought about. So the two approaches, natural products and synthesis, are complementary. Meanwhile, many of the methods that the synthetic chemist uses today are based on organometallic reagents – compounds that trace their ancestry directly back to Frankland's work of the 1840s.

The research effort and length of time that it takes to discover and develop a new natural product for drug use are graphically illustrated by one of the success stories of recent decades, taxol.[13]

In 1958, the US National Cancer Institute initiated a programme to screen 35,000 plant species for anticancer activity. Five years later, the Forest Service collected a large sample of bark from the Pacific yew tree, *Taxus brevifolia*, which was found to be active against tumours. There were, however, severe problems with extracting the active ingredient, which was present in only minute amounts, and initial test results on biological tissues were not conclusive. It was not until 1983 that phase I clinical trials on volunteer patients could begin, and these were delayed by allergic reactions that some of them developed to the solvent used to administer the drug. In the meantime, the chemists determined the structure of the active alkaloid $C_{47}H_{51}NO_{14}$, named taxol, and found that it contained a new type of ring system difficult to synthesize. Biochemical studies showed that taxol functions in the cell by a formerly unknown cellular mechanism, which greatly increased interest in the new substance.

In 1989, a 30 per cent response rate was found for patients with advanced ovarian cancer, but supply was still a problem. To market the drug fully would require harvesting all the bark from, and thus killing, the world's entire supply of the yew trees. A large Pacific yew takes 200 years to grow and it required the bark of six trees to treat one patient. Nevertheless, the NCI sought a commercial partner for further development. By the 1990s, the Bristol–Myers–Squibb giant had made taxol its leading research priority and it was launched as a commercial drug in 1993. Over 100 research groups worldwide investigated the possibility of an economic total synthesis. Although several syntheses were reported from 1997 onwards, the number of steps required and the low yields mean that a purely synthetic taxol is not a viable option. Instead, the method now used is a semisynthesis from another closely related (but pharmacologically inactive) natural product, 10-Deacetylbaccatin III. Not only does this occur in a commoner species, the European yew *Taxus baccata*, but it occurs in greater amounts and it is found in the needles, so the tree does not have to be killed to extract it. Taxol is now used against a range of tumours, although it is a palliative and not a cure. Research has now swung back to synthetic aspects – making many analogues of taxol in order to find some that show improved activity. All in all, it is a heart-warming though exorbitantly expensive tale. Bristol–Myers–Squibb's sales worldwide reached $1.2 billion in 1998. Since it became clear that the Pacific yews would not have to be chopped down, the only slight note of discord came when Bristol–Myers–Squibb were recently granted the name Taxol™ as a trade mark. This forces those who are unwilling to use the irritating superscript to refer to it by its international nongeneric name, which is Paclitaxel.

The way in which plants make substances like taxol is still almost as humbling to us today as it was to Humboldt or Bonpland. We know more about the process, but that does not necessarily make it less impressive.

The alkaloid chemist Sir Alan Battersby would often start his elegant lectures with a thought-provoking demonstration. In the centre of the bench he placed a large beaker of water, and dropped in a spoonful of a nitrogen-containing salt and a piece of solid carbon dioxide. He then switched on a lamp so that the gas bubbling up through the water was illuminated. The audience would watch in silence for a moment. Then, as if by an afterthought, he would remind them that he had forgotten the trace elements, and would drop in a piece of copper wire. Everyone would watch this uninspiring display for another few seconds. Nothing would happen, of course. Battersby would then remind the audience that, with these unpromising materials, the plants build not only themselves, but an array of hundreds of thousands of natural products including the sugars, aminoacids and all the other chemical building blocks that support the whole of animal and human life. Animals break down the starches and proteins of the plant tissues into these sugars and aminoacids and reassemble them in their own way, but they cannot make these building blocks from scratch. All animal life, whether herbivorous or carnivorous, depends on the plants.

To early man this must have seemed truly miraculous, and it still seems little short of that today. In a daily reversal of the deathly corruption that so preoccupied alchemists in the Middle Ages, the plant produces green shoots, fruits, the timber with which a family can build its house, drugs with which they can heal themselves. A plant seems immortal, for it retains the spark of life in its remotest leaf. But plants, despite the extraordinary achievement of being able to do this at all, utilize only a tiny percentage of the energy falling on them. For decades, there has been intensive research into how the plants, using the organometallic compound chlorophyll, carry out their photosynthesis, in the hunt for clues about how this process might be improved. Up until now, the results have been mediocre, but a breakthrough here could end all reliance on fossil fuels and totally transform society.

A 1975 Jewish prayerbook states: 'We study the seed and the cell, but the powers deep within them will always elude us.'[14] In the previous chapter we saw how Pasteur's experiments of the 1840s raised questions about the origin of life on earth that have still not been resolved. This is not the only area of enquiry in which the molecule poses us unresolved questions.

A new generation of neurologists is struggling with the problem of whether the functioning of the brain, and the genesis of consciousness, can be explained on mechanistic lines; is the brain a computer, as Turing thought, or is it something more?[15] There have been three main strands of thought about the fundamental nature of consciousness. The predominant strain in Western thought until fairly recently was that of Descartes, who said that mind and matter both exist, and are two separate aspects of the universe.

Few people now subscribe to the extreme view of Bishop Berkeley – that only mind exists. Samuel Johnson's stone-kicking experiment is just too persuasive. But more recently, with the growth of our knowledge of computing, there has been a tendency towards the materialistic hypothesis that only matter exists, and that consciousness is an 'Epiphenomenon', which comes about automatically when matter reaches the level of organizational complexity of the brain. Joseph Priestley would probably have been proud to be associated with such a view.

But a new picture is emerging: that perhaps matter and consciousness are not separate, but two sides of the same coin, and that the emergence of consciousness is intimately associated with the way in which the brain is made chemically. A silicon computer, even if it had the same level of complexity as the brain, would not be conscious.

In the model of the atom and the molecule that replaced the 'planetary' one used earlier in this book, the electrons are not tiny particles, but have a character somewhere between particles and waves. A particle has to be somewhere at a given instant of time, but a wave is in no one place. The equations that describe the behaviour of these wavelike electrons in even a small molecule cannot be solved exactly, only to a fair degree of approximation. We can model what the electrons do in small molecules reasonably well, but we do not fully understand them. As the molecule becomes larger, the uncertainties become greater. According to this picture, consciousness must derive from the fugitive, quantum nature of the electrons in the brain molecules, which cannot be corralled into definite pathways. We still do not fully understand the nature of even the small organic molecules that played such a major part in the emerging science of chemistry during the nineteenth century, and that this book has been about. Chemistry still has much to tell us about ourselves.

Glossary

Note. Terms marked* are not specifically referred to in the main text of the book, and are included in this glossary to assist explanation of some of the other terms.

Alkali   A strong soluble base (q.v.) – e.g. sodium or potassium hydroxides (caustic soda and potash).

Alkali metals   A family of related elements lithium, sodium, potassium, rubidium and caesium. Their atomic structure is characterized by an outer shell containing one electron, and their chemistry is dominated by their tendency to donate this electron to achieve a positive ion (q.v.) with the noble-gas (q.v.) structure. React with water to form alkalis.

Alkaloid   A nitrogen-containing plant (or animal) product. The presence of nitrogen usually results in basic properties. Originally called vegetable alkalis.

Aromatic compound   An organic compound containing one or more radicals derived from benzene or its close relatives.

Atom   The smallest particle of an element (q.v.).

*Atomic number**   The number of protons in the nucleus of an element, equal in the neutral atom to the number of electrons orbiting it. The atomic number determines the element. For the lightest element, hydrogen, it is 1. There is no known upper limit to atomic number, but beyond about 90 the atoms become progressively more unstable (radioactive). There are today no unknown elements of atomic number less than 100 meaning that all the spaces in the periodic table are filled. Not to be confused with atomic weight (q.v,), which in the early nineteenth century was often called atomic number.

Atomic weight   The relative weight of the atom of an element. In the nineteenth century it was most usual to express atomic weights relative to hydrogen = 1. Other scales were used: e.g. oxygen = 10 or oxygen = 100. Nowadays, atomic weights are expressed much more precisely with reference to a particular isotope (q.v.) standard, so that, in general, atomic weights are not whole numbers – e.g. hydrogen = 1.008 not 1. In the early nineteenth century, atomic weight was often referred to as atomic number.

Avogadro's number   The number of atoms in a gram-atom (q.v.), or the number of molecules in a gram-molecule (q.v.) – approximately 6×10^{23}, or 6 followed by 23 zeros. A gram-atom or gram-molecule of different gases occupies the same volume at the same temperature and pressure (22.4 litres at 25° and atmospheric pressure). The first experimental estimate of Avogadro's number was made by Loschmidt.

Base A compound that forms salts with acids (acid + base → salt + water).

Bond, chemical bond Linkage between two atoms in a molecule having direction in space. In the simplest cases, formed by the sharing of a pair of electrons (q.v.) between the two atoms. Double or triple bonds can be formed by the sharing of two or three pairs.

Calx Old term for an oxide, obtained by strongly heating (calcining) a metal in air.

Carbohydrates Important class of plant products usually made up of carbon, hydrogen and oxygen only. Examples are sugars and starches. In the common carbohydrates, the ratio of hydrogen to oxygen is the same as in water, hence the name.

Carbon Element of atomic number (q.v.) 6, the fundamental building block of organic compounds. It has four electrons in the outer shell, and its chemistry is dominated by the tendency to form covalent bonds (q.v.).

Catenation The ability of atoms of certain elements to form chains and rings by bonding together. Catenation is most strongly shown by carbon.

Chemical* (1) Any material substance. (2) Increasingly, if inaccurately, used for evil substances made by chemists.

Chemistry The science that investigates the different kinds of matter, including how elements combine to make compounds, and how and why chemical reactions happen.

Chirality Handedness.

Compound A pure chemical substance formed when two or more elements combine together, by forming covalent or ionic bonds.

Covalent bond* Bond formed by the sharing of two electrons between two atoms. Covalent bonds are directional, and responsible for the structure of the molecule. Characteristic of compounds between non-metals – e.g. carbon, oxygen, hydrogen, nitrogen.

Electron Subatomic particle carrying a negative charge, orbiting the nucleus and determining the chemistry of the element. About 2,000 times lighter than the proton.

Element Substance that cannot be broken down (by chemical means) into anything simpler. The atoms of a given element are all identical, except for isotopes (q.v.) and one element differs from another by virtue of its atomic number (q.v.).

Empirical formula The formula expressed as its simplest ratio – e.g. salt = $NaCl$, benzene = CH (or C_2H, as originally thought).

Equivalent weight The weight of one element combining with another, normally expressed relative to hydrogen (the lightest element) = 1. Thus the equivalent weight of oxygen in water, H_2O, is 8 (approximately). Equivalent weights can vary from one compound to another. Thus in hydrogen peroxide, H_2O_2, the equivalent weight of oxygen is 16. Equivalent weights could be found experimentally long before atomic weights were known.

Formula   The constitution of a compound, showing what it is made up of. *Molecular formula* = the actual number of atoms in a molecule – e.g. water = H_2O, benzene = C_6H_6, strychnine = $C_{21}H_{22}N_2O_2$.

Functional group   A group of atoms, especially in an organic molecule, attached to the carbon framework and imparting characteristic chemical properties. An example is the hydroxyl group, –O–H, in alcohols. See also radical.

Gram-atom, gram-molecule, gram-equivalent*   The atomic, molecular or equivalent weight of an element or compound expressed in grams – e.g. a gram-molecule of hydrogen (H_2) is 2 grams. A gram-atom or gram-molecule of any substance contains Avogadro's number (q.v.) of atoms or molecules. The gram-equivalent of every gas occupies approximately the same volume (22.4 litres) at atmospheric pressure. Now called a *mole** and formally defined as the amount of a substance containing Avogadro's number of particles, whether atoms, molecules or ions.

Halogens ('salt-formers')   A family of related elements chlorine, bromine and iodine, known by the early nineteenth century and having related chemical properties. The fourth halogen, fluorine, is so reactive that it could not be obtained as the free element until 1886, but its compounds were well known. The halogens have an outer shell of seven electrons and their chemistry is dominated by the tendency to gain the eighth electron needed for the noble-gas (q.v.) structure.

Hydrocarbon   A compound containing carbon and hydrogen only – e.g. methane, CH_4, benzene, C_6H_6. Since the electronegativities (electron-attracting powers) of carbon and hydrogen are approximately the same, the hydrocarbons can form the stable neutral molecular scaffolding to which functional groups (q.v.) can be attached.

Hydrogen   The lightest element (atomic number = 1) with an atom consisting of a single electron orbiting a single proton. Normal hydrogen gas is diatomic, H_2.

Inorganic chemistry   Formally, the chemistry of all elements except carbon. Originally, the chemistry of non-living things.

Ion   An electrically charged form of an element resulting from the addition or removal of electrons. Addition of one or more electrons gives a negatively charged ion (anion); removal of electrons gives a positively-charged ion (cation). *Ionic compound*   A compound made up of ions as opposed to molecules – e.g. sodium chloride.

Isomers   Two compounds with the same formula, differing in the way in which the atoms are joined together.

Isotopes*   Forms of an element having different numbers of neutrons in their nucleus. They have essentially identical chemical properties. The presence of mixtures of different isotopes explains why the atomic weights of many elements are not close to whole numbers – e.g that of chlorine =35.5.

Molecular weight   The weight of a molecule expressed as a ratio, normally in the nineteenth century by comparison with the atomic weight of hydrogen set as =

1. The molecular weight of a compound is the sum of the atomic weights of its constituent atoms.

Molecule   The smallest particle of a chemical compound that can exist, made of two or more atoms joined together.. The molecule of an element may consist of two or more identical atoms – e.g. Cl_2, S_8. Formerly often known as a compound atom.

Neutron*   Electrically neutral particle having approximately the same mass as a proton, and present in the nuclei of all elements except ordinary hydrogen. Variants of an element having differing numbers of neutrons in their nucleus are called isotopes (q.v.).

Nitrogen   Element of atomic number (q.v.) 7. Major constituent of air (as N_2). Many organic nitrogen compounds – e.g. alkaloids (q.v.) are basic (q.v.).

Noble gases (inert gases, rare gases)   Family of elements (Helium, Neon, Argon, Krypton and Xenon) in which the outermost electron shell of 2 or 8 electrons is full, and characterized by their chemical unreactivity. Not discovered until the 1890s.

Organic   (1) Originally, derived from living things. (2) Technically, compounds containing carbon. (3) Modern variation: overpriced product said not to contain any chemicals (q.v.), but, as all material substances are chemicals, obviously it does.

Organometallic compound   A compound containing a metal joined to an organic radical – e.g. diethylzinc, $C_2H_5.Zn-C_2H_5$, discovered by Frankland. Their discovery was a major factor leading to the abandonment of the Radical Theory.

Oxygen   Element of atomic number (q.v.) 8. Present in air (as O_2).

Phenyl   The radical, C_6H_5-, derived from benzene by removal of one hydrogen atom.

Phlogiston   In the eighteenth century, the essence thought to be given off by burning substances. Logically, must weigh less than nothing. Dismissed by Lavoisier as 'the spirit of minus oxygen'.

Proton   Subatomic particle carrying a positive charge, present in the nucleus of all atoms. The number of protons determines the atomic number (q.v.).

Pseudohalogen   A small, negatively charged radical (group of atoms) that mimics the chemical behaviour of the halogens – e.g. cyanate $(CNO)^-$.

Racemate   The form of a chiral substance in which equal amounts of right- and left-handed molecules are present. Racemates therefore show no optical activity. They are the invariable product when a substance with chiral molecules is made by conventional chemical methods.

Radical   Several atoms behaving as a group – e.g. phenyl (q.v.), methyl.

Radical theory   The descendant of Berzelius's dualistic theory of bonding current in the 1850s. Essentially, it lost out to the type theory (q.v.).

Type theory   Chemical theory based on Gerhardt's classifications and which led to the recognition of functional groups (q.v.) and later to the full structure theory of organic chemistry.

Valency   The combining power of an element, i.e. how many other atoms each of its atoms can combine with – e.g. H = 1, C = 4, hence CH_4. Many elements, especially metals, can exhibit more than one valency.

Vital force   Force or influence previously considered to act in addition to the known physical laws to govern the functioning of living organisms and their chemistry. Could be expressed either as a divine influence, and/or as the operation of hitherto unknown physical laws.

Notes

Introduction

1. This was F.R. Japp, who appears in Chapter 14.
2. Hooper 2001.
3. Rudofsky 1993: 18.

Chapter One

1. Hill and Powell 1934: 1: 471.
2. Ibid. 4: 27.
3. Ibid. 3: 398; 4: 237.
4. Bradley 1992: 12.
5. Liebig 1851: 36–7. See this book for other highly entertaining anecdotes of a similar nature collected by the irrepressible Liebig.
6. Ibid.
7. Partington 1962: 3: 226. I have been unable to trace the original refernce to this incident in the Evelyn diaries as cited by Partington.
8. Liebig 1851.
9. Ibid.
10. Partington, 1962: 2: 279.
11. Dobbin 1942: 83.
12. Greenberg 2000: 24.
13. Schaffer 2002; see also Haley 2002.
14. Greenberg 2000: 146.

Chapter Two

1. Albert 1987: 1.
2. Rees 1923: ch. 2.

3. Barrett 1905; Copeman 1967: 19.
4. Barrett 1905: 5.
5. King-Hele 2003: 54.
6. For a description of Vigani's cabinet at Cambridge and other contemporary apothecary's cabinets, see Haley 2002.
7. Haley 2002: 3.
8. Campbell 1747: 57–66.
9. Arnaud 1650: p. xviii, cited by Partington 1962: 3: 9.
10. French and Wear 1989: 191–221.
11. Rosenfeld 1997: 94.
12. Greenberg 2000: 167.
13. Knight 1992.
14. Greenberg 2000: 164.
15. Partington 1962: 2: 11.
16. Holmes 1985: 25.
17. Ibid. 320.
18. Liebig 1851: 45.

Chapter Three

1. Rosengarten 1959; Huayyin 1942.
2. Macpherson 1812; Furber 1951; Chaudhuri 1978.
3. Kochhar 1992.
4. Husemann 1857.
5. Humboldt and Bonpland 1808: p. ii.
6. Ibid. 5.
7. Ibid. 5.
8. Smeaton 1962.
9. Partington 1962: 3: 536. Other sources, e.g. Smeaton, do not

support this dramatic version of Fourcroy's demise on the same day that he was made a count.

10. Hamilton 2002: 66–7.
11. Humboldt and Bonpland 1808: 5.
12. Davy 1836: 1. 226–7.
13. Dobbin 1931: 247.
14. Ibid. 219.
15. Partington 1962: 3: 234.
16. Dobbin 1931: 261.
17. Serteuner 1805–11.
18. Christison 1885 1: 238.
19. Smeaton 1962: 163.
20. Pelletier and Caventou 1819.
21. Dobbin 1931.

Chapter Four

1. Mellor 1922: 1, ch. 1.
2. Bradley 1992: 22.
3. Mellor 1922: 79.
4. Greenaway 1966; Nye 1996: ch. 2.
5. Gangopadhyaya 1980: 30.
6. Greenaway 1966: ch. 7.
7. Ibid. 138, Rocke 1984: 21–35.
8. Partington 1962: 4: 154.
9. Rocke 1984: 38.
10. Schütt 1992: 120.
11. Liebig 1851: 102.
12. Rocke 1984: 37.
13. Gangopadhyaya 1980: 302.
14. Bradley 1992: 24.
15. Brock 1967: 1.
16. Knight and Kragh 1998: 34.
17. Anon. 1929.
18. Meldrum 1904.

Chapter Five

1. Volhard 1909; Heuss 1949; Brock 1997.
2. Hofmann 1876: 10.
3. Brock 1997: 177.

4. Mulder 1847 (emphasis in original).
5. Brock 1997: 21.
6. Ibid. 129.
7. Ibid. 24.
8. Christison 1885: 1: 239.
9. Gendron 1961: 134.
10. Hausen 1969: 48.
11. Browne 1938.
12. Jorpes 1966: 32.
13. Wöhler 1844.
14. Partington 1962: 4: 284.
15. Brock 1997: 45 ff.
16. Browne 1938.
17. Ramsay 1981: 48.
18. Brock 1997: 74.
19. Ibid. 186.

Chapter Six

1. Jorpes 1966; Melhado 1980.
2. Jorpes 1966: 71.
3. Söderbaum 1913: 3: 87–95.
4. Melhado 1980: 148.
5. Partington 1962: 4: 163.
6. Jorpes 1966: 34.
7. Priestley 1767.
8. Figuier 1867: 653; Pattison 1986.
9. Kahlbaum and Darbishire 1899: 31–3. Some continental newspapers were claiming that Faraday had exhibited the insects at a meeting of the Royal Institution.
10. Knight 1992: 65.
11. Partington 1962: 4: 162.
12. Dobbin 1942: 9.
13. Caven and Cranston 1928.
14. Ramsay 1981: 50.
15. Brock 1967: 142.
16. Larry Holmes, quoted by Brock 1997: 92.

Chapter Seven

1. Rocke 1984: 35.

2. Novitski 1992: 14.
3. Partington 1962: 4: 162.
4. Ibid. 211.
5. Bradley 1992: 45.

Chapter Eight

1. Watson 1817: 28.
2. Haley 2002.
3. Priestley 1777.
4. Holt 1931: 108, 124.
5. Hill and Powell 1934: 4: 238 nn.
6. King-Hele 2003: 36.
7. Prout 1855.
8. Melhado 1980: 119.
9. Hofmann 1876: 21.
10. Brock 1997: 203.
11. Cohen and Cohen 1996. All the quotations on the next few pages are from this paper.
12. Berzelius 1847: 4: 1–7.
13. Liebig 1843: 7.
14. Liebig 1867: 16.
15. Carpenter 1850.
16. Ibid (emphasis added).
17. Ibid.
18. Carpenter 1865.
19. Brock 1997: 327.

7. George Sand, the French novelist, describes how, when she was staying at a monastery on Majorca with Frederic Chopin in the winter of 1838–9, the Carthusian monk acting as pharmacist sold them some benzoin resin to counteract the abominable stink from the stove in their cell, which had been mortared with animal manure (Sand 1977: 170).
8. Kopp 1891.
9. Mitscherlich 1834.
10. Schütt 1992: 143.
11. Ibid. 144.
12. Partington 1962: 4: 331.
13. John Read in Perkin centenary Celebration Committee, 1958: 11.
14. Letters from Faraday to Grove, 1845, Faraday to Schonbein 1852 and Schoenbein to Liebig, 1853, given in Kahlbaum and Darbishire 1899: 206, 209. It should be noted though that around this time Faraday was suffering from mental deterioration and poor memory generally, which was a constant theme of his letters.

Chapter Nine

1. Hamilton 2002.
2. Williams 1965: 107.
3. Martin 1932: 1: 203.
4. Faraday 1825.
5. Six carbon atoms of weight 12 = 72, six hydrogens of weight 1 = 6, total 78, which is 39 times the weight of the hydrogen molecule H_2, weight 2.
6. Hausen 1969: 48. No chemist would invite a visitor to smell benzene today, because of its toxicity.

Chapter Ten

1. Christiansen 1994.
2. Tiffeneau 1925: 45.
3. Grimaux and Gerhardt 1900: 222.
4. Chaigneau 1984.
5. Grimaux and Gerhardt 1900: 32.
6. De Chazelles 1890: 160.
7. Rocke 2001: 49.
8. Kapoor 1970.
9. Laurent 1837a: 136.
10. Williamson 1855.
11. Grimaux and Gerhardt 1900: 42–3.
12. Ibid. 56.

13. Ibid. 81.
14. Ibid. 184. This is a retranslation from the French-language version given by Grimaux and Gerhardt; the original letter was in English, which Gerhardt also mastered.
15. Ibid. 226. This was not Paul Sabatier, the well-known Nobel prizewinner in chemistry, who was not born until 1854.
16. Dumas 1834.
17. Partington 1962: 4: 362.
18. Berzelius and Pelouze 1838.
19. Laurent 1837a: 139. Laurent 1855 (at the time of writing the reprint of these collected translations is still forthcoming).
20. Partington 1962: 4: 381.
21. Laurent 1837a: 139 (emphasis added).
22. Laurent 1837b: 326.
23. Wöhler 1840.
24. Dumas 1838.
25. Partington 1962: 4: 388.
26. Berzelius and Pelouze 1838.
27. Tiffeneau 1918: 28.
28. William Gregory, letter to Gerhardt, 18 July 1845, quoted at Grimeaux and Gerhardt 1900: 100 and at Novitski 1992: 69.
29. Fremy 1850.
30. Original citation mislaid (probably in the voluminous Berzelius–Liebig or Berzelius–Wöhler correspondence). In other places Berzelius is careful to emphasize the need for careful scientific method; organic chemistry was his blind spot.
31. Partington 1962: 4: 374.
32. Rocke 1987.
33. Tiffeneau 1925: 38.
34. Tiffeneau 1918: 173.
35. Ibid. 35. It seems that Gerhardt did not fully take the admonition on board; see also ibid. 123.
36. Ibid. 151.
37. Grimaux and Gerhardt 1900: p. ii.
38. Tiffeneau 1918: 178, 198.
39. Grimaux and Gerhardt 1900: 64.
40. Tiffeneau 1925: 24–6.
41. Grimaux and Gerhardt 1900: introduction.
42. Tiffeneau 1925: 131.
43. deMilt 1951.
44. Williamson 1855.
45. deMilt 1951.

Chapter Eleven

1. Herschel 1831: 299.
2. Jorpes 1966: 99.
3. Volhard 1909: 134–6, 147.
4. Garfield 2000; Perkin Centenary Celebration Committee 1958.
5. Russell 1996.
6. Palmer 1965: 27.
7. Russell 1971: 83.
8. Hofmann 1865.
9. Liebig 1851: 160 (almost identical wording appears in the 1844 edition).
10. Russell 1996: 79.
11. Frankland 1852. This is a highly simplified and condensed account of his experiments. There was also the intermediate product ethyl zinc iodide, C_2H_5–Zn–I.
12. Couper 1858.
13. Russell 1971: 79.
14. Couper 1858 (emphasis added).
15. Anschütz 1929.
16. Brock 1967: 7.
17. Brodie 1868. This paper would be incomprehensible without extended study. A critical review is given by Brock (1967: 31–90), although few people could follow even this

admirable simplification without the aid of amphetamines and a wet towel.
18. Brock 1967: 48.

Chapter Twelve

1. Leicester and Glickstein 1952: 231–8.
2. Nye 1996: 34.
3. Palmer 1965: 14.
4. Gaudin 1873.
5. Partington 1962: 4.
6. Novitski 1992: 14.
7. Tiffeneau 1918.
8. Palmer 1965: 28.
9. Strathern 2000.
10. Cannizzaro 1858. English-language translation available at http://dbhs.wvusd.k12.ca.us/Chem-History/Cannizzaro.html.
11. Tilden 1912.
12. Ihde 1961. A detailed record is given by Anschütz 1929: 1: 671–91.
13. Tilden 1912.
14. Cannizzaro 1858.
15. Meldrum 1904; Bradley 1992: chs 9–10.
16. Palmer 1965: 29.
17. Tilden 1912.

Chapter Thirteen

1. Baring-Gould 1937.
2. Benfey 1958. All subsequent verbatim quotations of Kekulé in this chapter come from this source.
3. Hill and Powell 1934: 1: 471.
4. Wotiz 1993: 263.
5. Anschütz 1929.
6. Wotiz and Rudofsky 1983
7. Anschütz 1929: 1: 38.
8. Palmer 1965: 2.
9. Russell 1971: ch. VI.

10. Frankland 1902.
11. Palmer 1965: 43.
12. Japp 1901.
13. Russell 1971.
14. Ibid. 121 (table IV)
15. Anschütz 1929.
16. Brooks 1998a.
17. Brooks 1998b.
18. Kuznetsov and Shamin 1993.
19. Rocke 1981.
20. USSR Academy of Sciences 1961.
21. Kekulé 1858.
22. Dobbin 1942: 45.
23. See e.g. Wotiz and Rudofsky 1993; Schwartz 1993; Rothenberg 1993.
24. Rudofsky and Wotiz 1988.
25. Partington 1962: 4: 555.
26. For a discussion, see Brooke 1993.
27. Noe and Bader 1993.
28. The term Loschmidt's number is also used to denote the number of molecules in a cubic centimetre of gas, which is 22,400 times smaller.
29. Loschmidt 1861. See also Loschmidt 1913, which is an edited reprint.
30. Noe and Bader 1993.
31. Anschütz 1929: 1: 305.
32. Rocke 1987.
33. Rocke 1993: 333.
34. Wotiz and Rudofsky 1989.
35. Štrbáňová and Janko 1993.

Chapter Fourteen

1. Rocke 1993.
2. Rocke 2001: 378, illustration. Some have suggested that Wurtz's opening sally was not naked chauvinism, but rather a call to arms to his countrymen not to neglect a subject that had its origins in France.
3. Ramsay 1881: 62.

4. Rocke 1993: 325.
5. Ibid . 327.
6. Gardner 1967; Eliel and Wilen 1994.
7. Debré 1994: 47; Nicolle 1961: 14.
8. Ramsay 1981: 47.
9. See references cited in Partington, 1962, IV; 386.
10. Tiffeneau 1918.
11. Schütt 1992: 110.
12. Palmer 1965: 39.
13. Japp 1898. See also ibid. 495–6 (letter by Karl Pearson), ibid. 592–3 (letter by Herbert Spencer), and ibid. 616–18 (letter by Japp).
14. Herbert Spencer, in ibid.
15. The latest research apparently shows that the inherent chirality of the universe's subatomic particles is far too weak ever to have induced chirality in molecules.

Chapter Fifteen

1. Wotiz (1993) is the mangled printed version of the symposium. Some contributions were withdrawn and the original publishers, Springer Verlag, rejected the book. I had hoped to obtain the editor's inside story of the disputes, but unfortunately Dr Wotiz was killed in a car accident before we could get together.
2. Mitscherlich 1965.
3. Sponsel and Rathsmann-Sponsel (2000).
4. Katz and Acton 1973.
5. Schwarz 1958: 272.
6. Prof. C.W. Rees, lecture presentation (1990s).
7. Williams 1990; Robinson 1976.
8. Robinson 1931.
9. Kröhnke 1952.
10. Thackray and Myers 2000.
11. Sir Derek Barton, lecture presentation (1970s).
12. Watson 1968: 29.
13. Goodman and Walsh 2001.
14. Cohen and Cohen 1996.
15. Tudge 2003; Penrose 1989.

Bibliography

Albert, Adrien (1987). *Xenobiosis; Food, Drugs and Poisons in the Human Body.* London, Chapman & Hall

Anon. (1929). *Centenaire de Marcelin Berthelot.* Paris

Anschütz, Richard (1929). *August Kekulé.* 2 vols. Berlin, Verlag Chemie

Arnaud, E.R. (1650). Introduction à la chymie, ou la vraye physique, Lyon, Claude Prost

Baring-Gould, S. (1937). 'The Murder of the Countess Goerlitz', in *The Fifty Strangest Stories Ever Told.* London, Odhams

Barrett, C.R.B. (1905). *The History of the Society of Apothecaries of London.* London, Elliott Stock

Benfey, O. Theodor (1958). 'August Kekulé and the Birth of the Structural Theory of Organic Chemistry in 1858', *Journal of Chemical Education*, 35: 21–3

Berzelius, J.J. (1847). *Lehrbuch der Chemie.* 4 vols. Dresden and Leipzig, Arnoldische Buchhandlung

Berzelius and Pelouze (1838). Untitled letter, in *Comptes Rendus*, 6, 629–48

Bradley, John (1992). *Before and after Cannizzaro.* Caithness, Whittles Publishing Services

Brock, W.H. (1967) (ed.). *The Atomic Debates; Brodie and the Rejection of the Atomic Theory.* Leicester, Leicester University Press

Brock, William H. (1997). *Justus von Liebig, the Chemical Gatekeeper.* Cambridge, Cambridge University Press

Brodie, B.C., Bart. (1868). 'The Calculus of Chemical Operations; being a Method for the Investigation, by means of Symbols, of the Laws of the Distribution of Weight in Chemical Change – Part I. On the Construction of Chemical Symbols', *Journal of the Chemical Society*, 21: 367–466

Brooke, John H. (1993). 'Doing down the Frenchies: How Much Credit Should Kekulé Have Given?', in Wotiz 1993: 59–76

Brooks, Nathan M. (1998a). 'The Evolution of Chemistry in Russia during the Eighteenth and Nineteenth Centuries', in Knight and Kragh 1998: 163–76

—— (1998b). 'Alexander Butlerov and the Professionalization of Science in Russia', *Russian Review*, 57: 10–24

Browne, C.A. (1938). 'The 'Banquet des Chimistes', Paris, April 22, 1867', *Journal of Chemical Education*, June: 253–9

Campbell, R. Esq. (1747). *The London Tradesman, being a Compendious View of All the Trades, Professions, Arts, both Liberal and mechanic, now practiced in the Cities of London and Westminster.* London, T. Gardner; facsimile edn by David & Charles, 1969

Cannizzaro, Stanislao (1858). 'Sunto di un corso di filosofia chimica fatto nella R. Università di Genova dal Prof. S. Cannizzaro', *Il nuovo cimento*, 7: 321–66

Carpenter, W.B. (1850). 'On the Mutual Relations of the Vital and Physical Forces', *Philosophical Transactions*, 727–57

—— (1865). 'On the Nature of the Physical and Vital Forces', in *The Correlation and Conservation of Forces: A Series of Expositions by Prof. Grove, Helmholz, Mayer, Faraday, Liebig and Carpenter, with an Introduction and Biographical Notice of the Chief Promotors of the New Views by E.L. Youmans.* New York, D. Appleton & Co.

Caven, R.M., and Cranston, J.A., (1928). *Symbols and Formulae in Chemistry: An Historical Study.* London and Glasgow, Blackie

Chaigneau, M. (1984). *Jean-Baptiste Dumas, sa vie, son œuvre 1800–1884.* Paris, Guy le Prat

Chaudhuri, K.N. (1978). *The Trading World of Asia and the East India Company 1660–1760.* Cambridge, Cambridge University Press

Christiansen, Rupert (1994). *Tales of the New Babylon: Paris 1869–1875.* London, Sinclair-Stevenson

Christison, Robert (1885). *The Life of Sir Robert Christison, Bart., Edited by his Sons,* 2 vols. Edinburgh, William Blackwood

Cohen, Paul S., and Cohen, Stephen M. (1996). 'Wöhler's Synthesis of Urea; How do the Textbooks Report It?', *Journal of Chemical Education*, 73: 883–6

Copeman, W.S.C. (1967). *The Worshipful Society of Apothecaries of London: A History 1617–1967.* Oxford, Pergamon Press

Couper, A. (1858). 'Sur une nouvelle théorie chimique', *Comptes rendus*, 1157–1160

Crosland, Maurice (1978). *Gay-Lussac, Scientist and Bourgeois.* Cambridge, Cambridge University Press

Davy, J. (1836). Memoirs of the Life of Sir Humphry Davy, Bart. 2 vols. London, Rees Orme, Brown, Green and Longman

Debré, Patrice (1994). *Louis Pasteur.* Baltimore, Johns Hopkins University Press

De Chazelles, R. (1890). *Jean-Baptiste Dumas.* Biographies du XIXᵉ Siècle. Paris, Bloud et Borrel

deMilt, Clara (1951). 'Auguste Laurent – Guide and Inspiration of Gerhardt', *Journal of Chemical Education*, 28: 198–204

Dobbin, L. (1931). *The Collected Papers of Carl Wilhelm Scheele.* London, G. Bell & Sons

—— (1942). *Occasional Fragments of Chemical History.* Edinburgh, printed for private circulation

Dumas, M.J. (1834). 'Recherches de chimie organique', *Annales de Chimie (Paris)*, 56: 113–54

Dumas, J.B. (1838). Untitled reply to a letter by Berzelius and Pelouze, *Comptes rendus*, 6: 646–8

Eliel, Ernest L., and Wilen, Samuel H. (1994). *Stereochemistry of Organic Compounds*. New York, Wiley

Faraday, M. (1825). 'Sur de nouveax composés de carbone et d'hydrogène, et sur d'autres produits obtenus pendant la décomposition de l'huile par la chaleur', *Annales de Chimie (Paris)*, 30: 269–91

Figuier, Guillaume Louis (1867). *Les Merveilles de la science*. Paris, Corbeil

Frankland, E. (1852). 'On a New Series of Organic Bodies Containing Metals', *Philosophical Transactions*, 142: 417–44

—— (1902). *Sketches from the Life of Edward Frankland, Concluded by his Two Daughters*. London, Spottiswoode & Co.

Fremy, M.E. (1850). 'Nouvelles observations sur les transformations que la chaleur fait éprouver aux acides tartarique et paratartarique', *Comptes rendus*, 31: 890–3

French, Roger, and Wear, Andrew (1989). *The Medical Revolution of the Seventeenth Century*. Cambridge, Cambridge University Press

Furber, Holden (1951). *John Company at Work: A Study of European Expansion in India in the Late 18th Century*. Cambridge, Cambridge University Press

Gangopadhyaya, M. (1980). *Indian Atomism: History and Sources*. Calcutta, K.P. Bagchi & Co.

Gardner, Martin (1967). *The Ambidextrous Universe*. Harmondsworth, Penguin

Garfield, Simon (2000). *Mauve: How one Man Invented a Colour that Changed the World*. London, Faber & Faber

Gaudin, M.A.A. (1873). *L'Architecture du monde des atomes*. Paris

Gendron, Val (1961). *The Dragon Tree: A Life of Alexander von Humboldt*. London, Longmans, Green

Goodman, Jordan, and Walsh, Vivien (2001). *The Story of Taxol: Nature and Politics in the Pursuit of an Anti-Cancer Drug*. Cambridge, Cambridge University Press

Greenaway, Frank (1966). *John Dalton and the Atom*, London, Heinemann

Greenberg, Arthur (2000). *A Chemical History Tour*. New York and Chichester, Wiley-Interscience

Grimaux, Édouard, and Gerhardt, Charles, Jr (1900). *Charles Gerhardt: Sa vie, son œuvre, sa correspondance*. Paris, Masson et Cie

Haley, C. (2002). *Boltheads and Crucibles: A Brief History of the 1702 Chair of Chemistry at Cambridge*. Cambridge, University of Cambridge Department of Chemistry

Hamilton, James (2002). *Faraday: The Life*. London, HarperCollins

Hausen, Josef (1969). *Was Nicht in Den Annalen Steht; Chemiker-Anekdoten*, Weinheim, Verlag Chemie

Herschel, J. (1831). *A Preliminary Discourse on the Study of Natural Philosphy*. 3rd edn. London, Longman, Rees, Orme, Brown & Taylor

Heuss, Theodor (1949). *Justus von Liebig; Vom Genius der Forschung*. Hamburg,

Hofmann & Campe Verlag

Hill, George Birkbeck, and Powell, L.F. (1934). *Boswell's Life of Johnson*. 6 vols. Oxford, Oxford University Press

Hofmann, August Wilhelm von (1865). *An Introduction to Modern Chemistry, Experimental and Theoretic*. London

—— (1876). *The Life-Work of Liebig: Faraday Lecture for 1875*. London, Macmillan

Holmes, Frederic Lawrence (1985). *Lavoisier and the Chemistry of Life*. Madison, Wis., University of Wisconsin Press

Holmes, F.L. (1993). 'The Evolution of the Chemical Industry; A Technological Perspective', in S.H. Mauskopf (ed.), *Chemical Sciences and the Modern World*. Philadelphia, University of Pennsylvania Press

Holt, Anne (1931). *A Life of Joseph Priestley*. London, Oxford University Press/ Humphrey Milford

Hooper, John (2001). 'The Kaiser's Kinks', *Guardian*, 24 May

Huayyin, S.A. (1942). *Arabia and the Far East: Their Commercial and Cultural Relations in Graeco-Roman and Irano-Arabian Times*. Cairo

Humboldt, Alexander, and Bonpland, Aimé (1808). *Plantae Aequinoctiales*. Vol. 1. Paris, A.F. Schoell

Husemann (1857). *Hufeland's Journal für praktischen Heilkunde*, p. 511

Ihde, Aaron J. (1961). 'The Karlruhe Congress: A Centennial Retrospect', *Journal of Chemical Education*, 38: 83–6

Japp, F.R. (1898). 'Stereochemistry and Vitalism', *Nature*, 57: 452–60

—— (1901). *The Kekulé Memorial Lecture; Memorial Lectures Delivered before the Chemical Society, no. VII (delivered December 1897)*. London, Gurney & Jackson

Jorpes, J. Erik (1966). *Jac. Berzelius, his Life and Work*. Stockholm: Almqvist & Wiksel

Kahlbaum, Georg. W.A., and Darbishire, Francis W. (1899). *The Letters of Faraday and Schoenbein*, Basle and London, Schwabe/Williams and Norgate

Kapoor, Satish C. (1970). *Dictionary of Scientific Biography*, ed. in chief Charles Coulston Gillespie, New York, American Council of Learned Societies/Scribner's

Katz, Thomas J., and Acton, Nancy (1973). 'Synthesis of Prismane', *Journal of the American Chemical Society*, 95: 2738–9

Kekulé, August (1858). 'Über die s.g. gepaarten Verbindungen und die Theorie der mehratomigen Radicale', *Liebigs Annalen*, 104: 129–50

King-Hele, Desmond (2003). *Charles Darwin's The Life of Erasmus Darwin*. Cambridge, Cambridge University Press

Knight, David (1992). *Humphry Davy: Science & Power*. Oxford, Blackwell

Knight, David, and Kragh, Helge (1998) (eds). *The Making of the Chemist: The Social History of Chemistry in Europe, 1789–1914*. Cambridge, Cambridge University Press

Kochhar, R.K. (1992). 'Science in British India, 1. Colonial Tool', *Current Science*, 63: 689–94

Kopp, H. (1891). *Woehler und Liebig, Untersuchungen über das Radikal der Benzoesäure, Ostwald's Klassiker der Exakten Wissenschaften*. Leipzig, Wilhelm Engelmann Verlag

Kröhnke, Franz (1952). *Hermann Leuchs, Chemische Berichte*, 85, pp. LV–LXXXIX

Kutznetsov, Vladimir I., and Shamin, Alexei, 'Butlerov and A. Kekulé's Contributions to the Formation of Structural Organic Chemistry', in Wotiz 1993: 211–22

Laurent, Auguste (1837a). 'Recherches diverses de chimie organique', *Annales de chimie et de physique*, 66: 136–54

—— (1837b). 'Suite de recherches diverses de chimie organique, cinquième partie', *Annales de chimie et de physique*, 66: 314–35

—— (1855). *Chemical Method, Notation, Classification and Nomenclature*, trans. William Odling, repr. forthcoming in the Thoemmes Library of Science, no. 2, Bristol, Thoemmes

Leicester, H.M., and Glickstein, H.S. (1952). *A Source Book in Chemistry 1400–1900*. New York, McGraw-Hill

Liebig, Justus von (1843). *Animal Chemistry, or Organic Chemistry and its Applications to Physiology and Pathology*, trans. W. Gregory. 2nd edn. London, Taylor & Walton

—— (1851). *Familiar Letters on Chemistry, in its Relations to Physiology, Dietetics, Agriculture, Commerce and Political Economy*, trans. W. Gregory. 3rd rev. and enlarged edn. London, Taylor, Walton & Maberly; republished as Vol. 6 in the series The Development of Chemistry 1789–1914, ed. David Knight, London, Routledge/Thoemmes Press, 1998

—— (1867). *The Development of Science among Nations*. Edinburgh, Edmonston & Douglas

Loschmidt, J. (1861). *Chemische Studien*. Vienna, Druck von Carl Gerold's Sohn

—— (1913). *Konstitutions-Formeln der Organischen Chemie in Graphischer Darstellung, herausgegeben von Richard Anschütz*. Leipzig, Verlag von Wilhelm Engelmann

Macpherson, David (1812). *The History of the European Commerce with India*. London, Longman, Hurst, Rees, Orme & Brown

Maddox, Brenda (2002). *Rosalind Franklin: The Dark Lady of DNA*. London, HarperCollins

Martin, Thomas (1932). *Faraday's Diary*. Vol. 1. London, G. Bell & Sons

Meldrum, A.N. (1904). *Avogadro and Dalton: The Standing in Chemistry of their Hypothesis*. Edinburgh, William Clay

Melhado, Evan M. (1980). *Jacob Berzelius: The Emergence of his Chemical System*. Stockholm, Almqvist & Wiksell

Mellor, J.W. (1922). *A Comprehensive Treatise on Inorganic and Theoretical Chemistry*, Vol. 1. London, Longmans & Co.

Mitscherlich, Alexander (1965). 'Kekulés Traum: Psychologische Betrachtung einer Chemischen Legende', *Die Zeit*, 17 Sept., pp. 19–20

Mitscherlich, E. (1834). 'Uber das Benzol und die Sauren der Oel- und Talgarten', *Liebigs Annalen*, 9: 39–56, including a commentary by J.v. Liebig

Mulder, G.J. (1847). *Liebig's Question to Mulder Tested by Morality*, trans. P.F.H. Fromberg. London and Edinburgh, Blackwood

Nicolle, Jacques (1961). *Louis Pasteur: The Story of his Major Discoveries*. New York, Basic Books

Noe, Christian, and Bader, Alfred (1993). 'Josef Loschmidt', in Wotiz 1993: 223–46

Novitski, Marya (1992). *Auguste Laurent and the Prehistory of Valence*. Chur, Switzerland and Philadelphia, Harwood Academic

Nye, Mary Jo (1996). *Before Big Science: The Pursuit of Modern Chemistry and Physics 1800–1940*. Cambridge, Mass.: Harvard University Press

Palmer, W.G. (1965). *A History of the Concept of Valency to 1930*. Cambridge, Cambridge University Press

Partington, J.R. (1962). *A History of Chemistry*. 5 vols. Macmillan

Pattison, F.L.M. (1986). 'The Clydesdale Experiments: An Early Attempt at Resuscitation', *Scottish Medical Journal*, 31: 50–2

Pelletier and Caventou (1819). 'Sur un nouvel alcali végétal (la strychnine) trouvé dans la fève de Saint-Ignace, la noix vomique, etc.', *Annales de chimie*, 10: 142–77

Penrose, Roger (1989). *The Emperor's New Mind*. Oxford, Oxford University Press

Perkin Centenary Celebration Committee (1958). *100 years of Synthetic Dyestuffs*. Oxford, Pergamon

Priestley, Joseph (1768). *A Familiar Introduction to the Study of Electricity*. London

—— (1777). *Disquisitions relating to Matter and Spirit, To Which is Added, the History of the Philosophical Doctrine Concerning the Origin of the Soul, and the Nature of Matter, with its Influence on Christianity, etc.* London

Prout, W. (1855). *Chemistry, Meteorology and the Function of Digestion, Considered with Reference to Natural Theology*. 4th edn. London, Henry G. Bohn

Ramsay, O.B. (1981). *Stereochemistry (Nobel Prize Topics in Chemistry)*. London, Heyden

Rees, J. Aubrey (1923). *The Worshipful Company of Grocers: An Historical Retrospect*. London, Chapman & Dodd

Robinson, R. (1931). *Proceedings of the Royal Society of London, Section A*, p. 431

Robinson, Sir Robert, (1976). *Memoirs of a Minor Prophet: 70 Years of Organic Chemistry*. Vol. 1., Amsterdam and Oxford, Elsevier

Rocke, Alan J. (1981). 'Kekulé, Butlerov and the Historiography of the Theory of Chemical Structure', *British Journal for the History of Science*, 14: 32–52

—— (1984). *Chemical Atomism in the Nineteenth Century: From Dalton to Cannizzaro*. Columbus, Oh., Ohio State University Press.

—— (1987). 'Kolbe versus the "Transcendental Chemists": The Emergence of Classical Organic Chemistry', *Ambix*, 34: 156–8

—— (1993). *The Quiet Revolution: Hermann Kolbe and the Science of Organic Chemistry*. Berkeley and Los Angeles, University of California Press

—— (2001). *Nationalizing Science: Adolphe Wurtz and the Battle for French Chemistry*. Cambridge, Mass., MIT Press

Rosenfeld, Louis (1997). *Four Centuries of Clinical Chemistry*. Amsterdam, Gordon & Breach

Rosengarten, Frederick, Jr. (1969). *The Book of Spices*, Wynnewood, Pa., Livingston Publishing Co.

Rothenberg, Albert (1993). 'Creative Homospatial and Janusian Processes in Kekulé's Discovery of the Structure of the Benzene Molecule', in Wotiz 1993: 285–312

Rudofsky, Susanna (1993). 'The Benzolfest', in Wotiz 1993: 9–20

Rudofsky, Susanna, and Wotiz, John H. (1988). 'Psychologists and the Dream Accounts of August Kekulé', *Ambix*, 35: 31–8

Russell, C.A. (1971). *A History of Valency*. Leicester, Leicester University Press

Russell, Colin A. (1996). *Edward Frankland: Chemistry, Controversy and Conspiracy in Victorian England*. Cambridge, Cambridge University Press

Sand, George (1977). *A Winter in Majorca*, trans. W. Kirkbride. Palma, Majorca, Mossèn Alcovar

Schaffer, Simon (2002). Vigani and the Beginnings of Cambridge Chemistry, Lecture Presentation, Cambridge.

Schofield, Robert E. (1997). *The Enlightenment of Joseph Priestley: A Study of his Life and Work from 1733 to 1773*. University Park, Pa., Pennsylvania State University Press

Schütt, Hans-Werner (1992). *Eilhard Mitscherlich, Prince of Prussian Chemistry*, trans. William E. Russwy, Washington, American Chemical Society and the Chemical Heritage Foundation

Schwarz, Robert (1958). *Aus Justus Liebigs und Friedrich Wöhlers Briefwechsel in den Jahren 1829–1873*. Weinheim, Verlag Chemie.

Schwartz, Wynn (1993). 'Problem Representation in Dreams', in Wotiz 1993: 277–84

Serteuner, F.W. (1805–11). *Trommsdorffs Journal of Pharmazie*, 13: 234; 14: 467; 20: 99

Smeaton, W.A. (1962). *Fourcroy, Chemist and Revolutionary, 1755–1809*. Cambridge, W. Heffer & Sons

Société Chimique de France (1916). 'Le Centenaire de Charles Gerhardt'. *Bulletin de la Société*, supplement

Söderbaum, H.G. (1913). *Jac. Berzelius Bref*. Uppsala, Almqvist & Wiksells

Sponsel, Rudolf, and Rathsmann-Sponsel, Irmgard (2000). *Kekulés Traum*, Internet Publikation für Allgemeine und Integrative Psychotherapie, http:www.sgipt.org/th_schul/pa/kek/pak_kek0.htm

Strathern, Paul (2000). *Mendeleyev's Dream: The Quest for the Elements*. London, Hamish Hamilton

Štrbáňovà, Soňa, and Janko, Jan (1993). 'Kekulé's Character in Light of his

Ennoblement', in Wotiz 1993: 211–22

Thackray, Arnold, and Myers, Minor, Jr. (2000). *Arnold O. Beckman: One Hundred Years of Excellence*. Philadelphia, Chemical Heritage Foundation

Tiffeneau, M., (1918) (ed.). *Correspondance de Charles Gerhardt*. Vol. 1. Paris, Masson et Cie

—— (1925) (ed.). *Correspondance de Charles Gerhardt*. Vol. 2. Paris, Masson et Cie

Tilden, Sir William A. (1912). *Chemical Society Cannizzaro Memorial Lecture*. London

Todd, Lord. and Cornforth, J.W. (1976). *Obituary Notices of Fellows of the Royal Society*, p. 415

Tudge, Colin (2003). *Guardian*, 30 January, p. 11

USSR Academy of Sciences (1961). *Centenary of the Theory of Chemical Structure*. Moscow, USSR Academy of Sciences

Volhard, Jacob (1909). *Justus von Liebig*. 2 vols. Leipzig, J.A. Barth Verlag

Watson, James D. (1968). *The Double Helix*. London, Weidenfeld & Nicolson

Watson, R. (1817). *Anecdotes of the Life of Richard Watson, Bishop of Llandaff*. London, Cadell & Davies

Williams, L. Pearce (1965). *Michael Faraday*. London, Chapman & Hall

Williams, Trevor I. (1990). *Robert Robinson, Chemist Extraordinary*. Oxford, Clarendon Press

Williamson, A. (unsigned, attributed) (1855). Obituary notice on Auguste Laurent, *Journal of the Chemical Society*, 149–57

Wöhler, F. (1840) (Writing as S.C.H. Windler). *Liebigs Annalen*, 33: 308–10

—— (1844). *Liebigs Annalen*, 52: 152–4

Wotiz, John H. (1993) (ed.). *The Kekulé Riddle: A Challenge for Chemists and Psychologists*. Clearwater, Fla., Cache River Press

Wotiz, John H., and Rudofsky, Susanna (1983). 'Kekulé or Kekule?', *Ambix* (Nov.), 133

—— (1989). 'Louis Pasteur, August Kekulé and the Franco-Prussian War', *Journal of Chemical Education*, 66: 34–6

—— (1993). 'Herr professor Doktor Kekulé: Why Dreams?', in Wotiz 1993: 247–76

Index

DATE DUE

GAYLORD

PRINTED IN U.S.A.